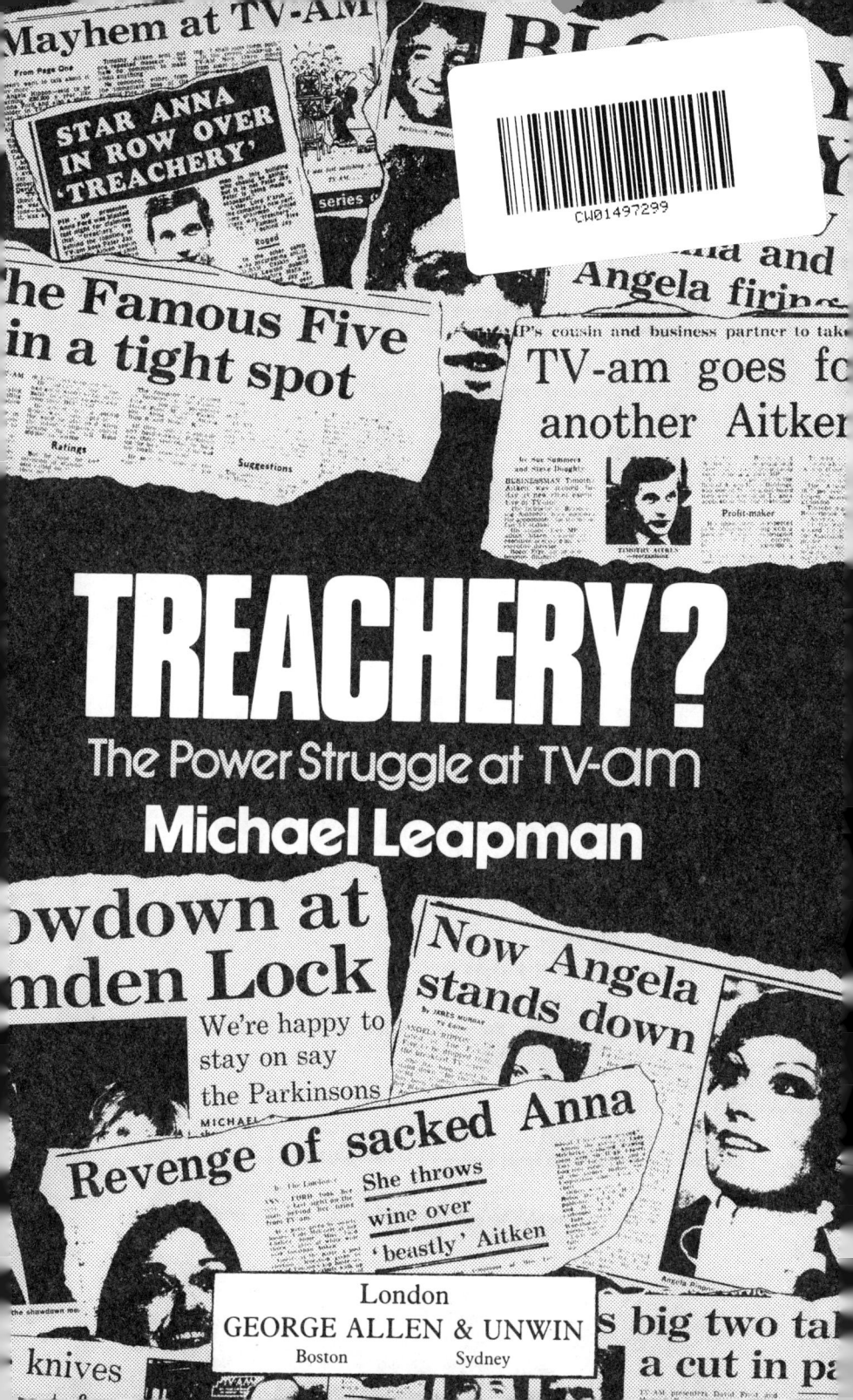

George Allen & Unwin (Publishers) Ltd,
40 Museum Street, London WC1A 1LU, UK

George Allen & Unwin (Publishers) Ltd,
Park Lane, Hemel Hempstead, Herts HP2 4TE, UK

Allen & Unwin Inc.,
9 Winchester Terrace, Winchester, Mass 01890, USA

George Allen & Unwin Australia Pty Ltd,
8 Napier Street, North Sydney, NSW 2060, Australia

First published in 1984

British Library Cataloguing in Publication Data

Leapman, Michael
    Treachery?: the power struggle at TV-am.
1. TV-AM 2. Television broadcasting—
Economic aspects—Great Britain
I. Title
384.55'43'0941   HE8700.9.G7
ISBN 0-04-791041-0

Set in 11 on 12 point Plantin by Grove Graphics, Tring, Hertfordshire
and printed in Great Britain by Biddles Ltd, Guildford, Surrey

*'There's been a great deal of treachery . . .
History will expose those who have been most
treacherous.'*

Anna Ford,
18th March, 1983.

Contents

A Note on Style and Sources

This is not a book about television, as I recognised before I had delved far into the research. It is a classic cloak-and-dagger boardroom melodrama, with one important difference: many of the leading characters are familiar faces from the TV screen. For that reason much of the action was played out on the front pages of the tabloid newspapers, instead of the inside pages of the Sunday business sections. But what lay behind those black headlines was, as I quickly discovered, even more intriguing than what they revealed.

It is a story about broken ideals and betrayed hopes. It concerns a group of well-meaning people who believed that, because they had been so successful at their craft of television, they could invent a new kind of company which would eliminate many of the failings they saw in the contemporary news media. Their venture foundered because they came up against an immutable City law: companies cannot operate without finance, and those who provide finance can also withdraw it.

A quality that stood out in the Famous Five and the other founders of TV-am was their sense of unity, of being a team not just on the field of battle but off it. They were a gang of firm friends who loved an excuse to crack open a case of champagne to celebrate their friendship. That has dictated a stylistic quirk of this book: the central characters are usually referred to by their first names, to sustain the impression of a cohesive group of intimates. The index of first names that follows will help. Where I have not been totally consistent in this it is mostly because of the surfeit of Michaels – Parkinson, Deakin and Rosenberg, all of them founder members of the gang. If anyone finds the device irritating, I apologise for it. At least I have spared you the well-known pet names of Frostie for David Frost and Parkie for Michael Parkinson. My wife Olga – for whose rigorous indexing and editing I am grateful – says the plethora of first names recalls those Peter-and-Jane readers for children; well, the Famous Five were, after all, a creation from children's literature.

The book is based on interviews with most of its leading characters. Since some wanted to speak in confidence it seemed politic not to identify any, but they know how grateful I am to them. My thanks, too, to all the others who have helped in numerous ways in the genesis and production of this book. For the interpretation of the facts I have garnered, and for the final dénouement, I accept sole responsibility.

Index of First Names

Many of the characters in the melodrama are referred to by their first names for much of the time. To assist readers, here is a check-list of the most common first names used, together with the appropriate surnames. The book is conventionally indexed by surnames at the back.

ANGELA Rippon, one of the Famous Five presenters.
ANNA Ford, another of the Five.
DAVID Frost, the Five's founding father.
DEREK Stevenson, director of sales (advertising).
DICK (Lord) Marsh, vice-chairman, later chairman.
ESTHER Rantzen, the first defector.
GEOFF Smith, director of operations.
JENNIE Bland, the token woman on the board.
JONATHAN Aitken MP, leader of the anti-Jay *putsch*.
KEITH Ewart, whose Wandsworth studio was to have been TV-am's headquarters.
MICHAEL Deakin, director of programmes.
MICHAEL Parkinson, one of the Famous Five.
MICHAEL Rosenberg, David's financial adviser.
NICK Elliott, the second defector.
PETER Jay, chairman and chief executive.
ROBERT Kee, last of the Famous Five.
TIMOTHY Aitken, Jonathan's cousin and financial minder.
TONY Wakeling, director of finance.

CHAPTER 1

The Man Who Came to Dinner

Smith Square is one of London's more civilised enclaves. A few minutes' walk from Parliament, dominated by the splendid baroque St John's Church, the square contains the Conservative Central Office, the headquarters of the Transport and General Workers' Union, and a cluster of eighteenth-century terraces occupied by a few of our wealthier MPs and national figures.

At night, apart from the occasional hurried steps of members rushing to the Commons for a division, the square is deserted and calm. Few strangers venture there; but any who had, shortly before midnight on Monday, 14 March, 1983, would have encountered a singular vision. Two men, both tall and lean, wearing dinner jackets with black bow ties, were walking slowly round the perimeter pavement making for nowhere in particular, engaged earnestly in conversation.

Had the stranger moved close enough to eavesdrop, he would have found the talk even and good-tempered. He would scarcely have guessed that one of the men had a few days earlier embarked on a determined effort to oust the other from his position as chairman and chief executive of Britain's newest and most controversial television company. Still less could he have determined from the calm, polite demeanour of the potential victim that he was entirely aware of the moves being made against him, and bent on resistance.

Ever since Oxford, twenty-five years earlier, Peter Jay had become accustomed to being called one of the most promising young men of his generation, and to some extent that promise had been fulfilled. In 1974 he was included by *Time* magazine in a list of the 150 people most likely to achieve leadership in Europe. (The list was not too prescient and Peter still finds the accolade embarrassing. Others in it included Shirley Williams, now in the political wilderness, and Nicholas Scott, still only a junior minister because he found himself on the wrong side in the Tory ideological schism. Only Prince Charles has so far stayed on course to fulfil the potential *Time* expected of them.) Two years later Peter was the subject of a revealing profile by Mark Amory in the *Sunday Times Magazine* headlined: 'The Cleverest Young Man in England?', only the question mark sparing deeper blushes.

Peter was born in 1937 into a left-wing intellectual family in Hampstead, North London. His father, Douglas Jay, was a Labour minister in the 1960s. Peter went to Winchester, by tradition the public school most devoted to matters of the mind. It was, he told Amory, 'an intense emotional experience'. He was head of the house and he confided:

'I remember for the only time in my life feeling that the responsibility was too great.'

Those agonising doubts about his capacity soon disappeared. He won a scholarship to Christ Church, Oxford, where some of his tutors began saying he was the cleverest person they had ever taught. After two defeats he was elected President of the Union, which he found 'enormously gratifying, because it got me back on track as a person who succeeds.' Forging ahead, he won a first-class degree in Politics, Philosophy and Economics. Soon, though, he slipped off the main track again when he failed in his bid for a fellowship at All Souls. 'It was a watershed in a big way because it was really the end of the phase of my life in which success, total success, was taken for granted.'

He had assumed he would make a career in Labour politics, like his father, so it was appropriate that at Oxford he should meet, fall in love with, and marry Margaret Callaghan, the daughter of the rising Labour politician who was later to become Chancellor of the Exchequer and then Prime Minister. But politics is not a career you can join fresh from university, so he went to the Civil Service instead. He was marked out as a high flyer notably early, rising to a Principal at the Treasury at the tender age of twenty-seven, in charge of the nation's education budget.

In 1967 William Rees-Mogg, the newly appointed editor of *The Times*, felt that Jay had exactly the kind of incandescent talent he needed to revitalise the newspaper, just purchased by the Canadian Roy Thomson. Rees-Mogg believed firmly in the power of the intellect. He was, by calculated design, a thinker rather than a man of action. For that reason, he and Peter were attracted to each other. Both believed there were few problems that could not be solved by a mixture of rational thought and goodwill. In their world there was no need for unseemly clashes, trials of strength or personal vindictiveness. Peter's philosophy differed from that of Rees-Mogg only in that, where his editor positively favoured thought over action, he placed equal emphasis on both. Peter has always believed that he is uniquely qualified to translate his beliefs into dynamic purpose. Like many, he found dynamism was a quality not much valued in the Civil Service. So when Rees-Mogg gave him the chance to switch careers and join *The Times* as economics editor, he agreed with alacrity. He was thirty years old at the time.

Just before he left he went to the annual Treasury staff dance at the Mecca Ballroom in Tottenham Court Road. It was not something he enjoyed – indeed, every year he dreaded it – but he would have been thought snooty had he declined to attend and it was, after all, the last year he would qualify. There he fell into conversation with Sir Alec Cairncross, head of the Government's Economic Service, and told him he wondered whether his career switch was the right decision.

'My boy,' Cairncross replied, 'What you must remember is that in life there are no wrong decisions: just decisions.' Peter, delighted, regarded this as the most liberating advice he had ever been given, and is wont to quote it when invited to address gatherings of young people. Others see it as positively

Mephistophelean in its irresponsibility, perhaps the root of some of his subsequent mishaps.

On *The Times* Peter became known as the writer of close-woven economic analyses comprehensible only to an elite. Certainly his former colleagues at the Treasury admired them and they were regarded as important by the international financial community, but they gained him no popular renown. And renown was what he sought – or, to be more precise, what he believed he was destined to achieve.

In 1970, thinking the time now ripe for his entry into politics, he sought the Labour nomination for Islington. He did not win it and the experience rather put him off. All that scrabbling at the grass roots was perhaps not for him after all. At *The Times*, however, he did go through a surprising spell as a trade union activist, encouraging members of the National Union of Journalists, by persuasive argument, to hold out for wage increases that would match the growing rate of inflation. As a result of his advocacy the union chapel threw out its moderate leaders and won a substantial negotiating success.

An opportunity to broaden his audience came when he was invited by Yorkshire Television to present a documentary programme about the life of Vic Feather, the Secretary-General of the Trades Union Congress. Although he was working with a brilliant young film-maker named Michael Deakin, his TV debut was not a success. Most of it, indeed, ended on the cutting-room floor.

But it was a start and viewers were not for long to be deprived of Peter's striking profile, his quizzical, full-lipped countenance. In 1972 he was asked to become the presenter of London Weekend Television's new Sunday programme 'Weekend World', which dealt with politics and current affairs in a more sober way than almost any other on television, mainly through the medium of long interviews with public figures, which Peter would conduct. His producer was John Birt and the pair used to enjoy debating the ethics of television and in particular what they saw as the inadequacies of current affairs programmes. Out of these discussions came a series of articles in *The Times* in 1975. Published under their joint byline, these became the subject of intense debate among television professionals, who more than most have a penchant for critical analysis of their work. The Jay-Birt thesis, as it came to be known, postulated that the makers of news and current affairs programmes had become too entranced with the visual image, that their selection of stories and the way they reported them were too dependent on whether they had any good action film to sustain the viewer's interest. This, wrote Jay and Birt, produced a 'bias against understanding.' The viewer would see yards of film of people firing guns into the middle distance, with no adequate explanation of the underlying causes of the conflict. In 'Weekend World', partly because they had few film crews to deploy, they relied much more on studio interviews, explanation and discussion. Cynics in television – they are legion – believed that the Jay-Birt thesis was simply a way of justifying expediency by dressing it up as a fancy theory. Yet it did help establish Peter as a thoughtful professional. In November 1975 he presented a closely argued twenty-four-page essay to Lord Annan's committee on the future of broadcasting.

Peter was able to combine his weekly screen appearances with his work at *The Times* but his next career advance meant leaving journalism for a while. In 1977 the Labour Foreign Secretary, Anthony Crosland, died in office. James Callaghan was now Prime Minister and Peter, his son-in-law, had easy access to him. A friend, Dr David Owen, was a junior minister at the Foreign Office and Peter was among those who suggested he should be promoted to the Cabinet post. Although strictly Owen was nowhere near senior enough for such advancement, the idea had the attraction for the Prime Minister of not involving a major Cabinet reshuffle, and the appointment was made.

Not long afterwards Dr Owen was able to repay Peter for his support. After a trip to America he decided Britain needed a new Washington ambassador and suggested to Mr Callaghan that he offer the post to his son-in-law instead of a career diplomat. Ignoring the inevitable allegations of nepotism he did so. Peter held the job for two years until the Conservatives came to power and quickly replaced him.

On balance his spell as an ambassador could be counted a diplomatic success, yet it was an uncomfortable period for two reasons. Professionally, he had serious clashes with his information staff, angered at his attempt to halve their number, and with John Robinson, deputy ambassador for much of his stay. And domestically, he split from Margaret, who fell into a close friendship, well documented in the gossip columns, with Carl Bernstein, one of the two *Washington Post* reporters responsible for uncovering the Watergate scandal. Bernstein's wife Nora Ephron later wrote a thinly-disguised *roman-à-clef* about the affair, airing the scandal anew. And in 1984, when the embassy nanny claimed Peter was the father of her four-year-old son, he admitted it was possible.

Sometimes Peter wondered how a person so devotedly rational as he could get involved in such private and public conflicts. One of his firmest beliefs was that people of first-rate intelligence should be able to argue out their differences calmly and reach acceptable accommodations. If Washington provided the first strong evidence against that view, his subsequent experiences were to discredit it totally.

He stayed in America after quitting the embassy in 1979, working desultorily on a research project and as an adviser to an oil company. It was there early the following year that David Frost – TV star, entrepreneur and constant transatlantic traveller – telephoned and invited him to participate in a new endeavour he and some associates were launching. In January the Independent Broadcasting Authority, in guarded and non-committal terms, had invited tenders for a new national television station devoted to broadcasting roughly between 6 and 9 a.m. Breakfast television was popular in the United States and other countries but had scarcely been tried in Britain, chiefly because existing agreements with the technical unions meant that they had to be paid overtime at extravagant rates for working in the small hours of the morning. By the terms of the IBA's prospectus the proposed breakfast programmes would consist of news and current affairs. The challenge attracted people in that area of television who were dissatisfied with the way the two existing networks handled the reporting and analysis of the news. Eight groups were being formed to apply for the franchise and Peter had already been approached by a number of them.

Peter seemed an ideal choice to play a leading role in such a project, for he possessed a unique blend of appropriate qualities. As a former ambassador, he had automatically become one of the Great and Good, a tribe whose participation was deemed essential to win approval from the IBA. (Made up largely of lay people with little expertise in television, the IBA was impressed, like most of us, with rank.) He combined this with experience as a presenter on a serious – if sparsely watched – current affairs programme, and with a reputation as a thinker on the issues behind the screen, derived from those *Times* articles.

There was also an excellent precedent for a former Washington ambassador running a TV company. John Freeman, once an on-screen interviewer, had returned from his stint in the American capital a decade earlier to take the reins at the then troubled London Weekend Television. By common consent, he had done the job marvellously, piloting the company to its present position as one of the most prosperous and effective in the independent network.

As it happened, David Frost had been in at the birth of London Weekend as well, and he had created it in much the same way as he was now putting together his breakfast company. He had trawled some of the most glittering names in television and the arts and ensured their participation, creating an irresistibly star-studded package to present to the IBA. Many thought this emphasis on famous faces rather than solid business acumen had been responsible for London Weekend's early floundering but David did not agree. Although he no longer had any working association with the company, he claimed as much credit for its ultimate success as blame for its initial failure – or, as he preferred to put it, its teething troubles. Certainly he did not regard that experience as a deterrent to having a try at the new breakfast franchise.

Although David's was not the first invitation to Peter to join a breakfast consortium, Peter was attracted to it. The first reason was personal: he had entertained David on several occasions at the Washington embassy and had grown rather fond of him. More important, however, was the consideration that the consortium and its ideas were scarcely yet formed. The other hopefuls had approached him after they had already done a lot of work on their structure and basic programme philosophy. Two of the groups had been in existence since even before the

IBA invited applications for the franchise. David's outfit, though, was still no more than a notion tossed between him and three others: Michael Rosenberg, his business associate; Richard Marsh, the former Cabinet Minister and ex-chairman of British Rail; and Michael Deakin, the man who had nursed Peter through his fumbling TV debut more than ten years earlier. Deakin now produced most of David's programmes for Yorkshire Television, as well as being co-author of many of his popular books. Peter had far fewer current commitments than any of these, so he could have the top executive position in the new grouping and play a creative role in its formation. He loved being in charge. Early in 1980, over lunch at La Caravelle, a restaurant in New York, he said yes.

It was a tremendously exciting year. David and Peter's joint cachet, combined with generous financial bait, lured as many of the top names in television to their ranks as they thought they needed. Scarcely anyone turned them down. Michael Parkinson, host of a popular BBC talk show, was the first to be recruited. He was followed by Robert Kee, the veteran current affairs specialist, and the BBC news reader Angela Rippon, famous for showing her legs on the Morecambe and Wise show. The prize catch was Esther Rantzen, producer and presenter of 'That's Life', the consumer watchdog and public participation programme that regularly won higher ratings than anything else on the BBC. On the production side they secured Nick Elliott, a promising young executive from London Weekend.

It was not just the high salaries and the six-month annual holiday that attracted the stars. More important was that they were going to be able to buy equity in the company at about half the price outsiders would have to pay. What is more, they were told the station would be run along the lines of United Artists in pre-war Hollywood, with the presenters having a good deal of say about the content of the programmes they presented. That had been Michael Deakin's idea, formed as he sat in the garden of David Frost's Hollywood home, looking out over Los Angeles.

To appreciate just how seductive a proposition it was, it is necessary to understand something of the constant tension in

TV between people who appear on the screen and the production staff who actually get the programmes on the air. The presenters are sarcastically dismissed as 'magic puppets' by the backroom people, who envy the inflated sums they are paid for doing little more than reading off an autocue. In return, the presenters resent being treated by production staff as dimwits with no constructive ideas of their own. They feel highly insecure, and with reason, for the price of their high salaries is the knowledge that at any time they could fall from grace, losing their crowd appeal and their earning power. They know, too, that it is mainly the competence of the production crew that prevents this happening – and by the same token it is the staff's fault if things go dreadfully wrong. They are quite at the mercy of others. Thus, the prospect of a piece of equity, coupled with the notion that they would now be less dependent on the people behind the cameras, was irresistible. The famous faces did not resist.

During the spring someone suggested adding to their number by approaching Anna Ford, an attractive news reader for Independent Television News. Peter was unenthusiastic and the matter was not initially pursued. Then in August he was invited to chair a discussion at the International Television Festival in Edinburgh. Coinciding with the long-established Edinburgh Festival, this newer event gave TV people the chance to indulge in a favourite pastime, agonising about the ethics and standards of their calling in self-indulgent rap sessions. Peter was in charge of a panel deliberating on current affairs TV. The one just before it was about women in the media, and Anna Ford was in the chair. Peter was taken aback when he learned her identity, and now realised that when her name had been mentioned to him in the spring he had thought she was someone else (to be precise, Sue Lawley of the BBC's 'Nationwide'). He had been in Washington during the period of Anna's rise to celebrity, and as a result her face was less familiar to him than to the millions of regular viewers of 'News at Ten'. Sitting through her session he found himself increasingly impressed by Anna's ability to handle herself and the other speakers, and by her cool, precise intelligence. Returning to London, he lobbied strongly among his new partners for an approach to be made. Hers was the last of the famous faces to be hoisted on board.

By the time she arrived the gang were already beyond the first stage in their bid for the franchise – the drafting of the application document. This had to be in the hands of the IBA by mid-May and had been drawn up during a hectic but stimulating Bank Holiday weekend at David Frost's house in Egerton Crescent, near Harrod's. First they needed two key elements. They had to identify the studio they would use if they won the franchise. After asking around, they hit on a modern facility in Wandsworth, South London, built from his own resources by Keith Ewart, a former fashion photographer. He had agreed that they could tell the IBA they planned to operate from his place, which had the advantage of adjoining waste land that could be used for supporting offices.

More important, they needed financial backing. Michael Rosenberg arranged the major loan through Barclay's Merchant Bank and they in turn canvassed for shareholders, among them the Rothschild Investment Trust. David Frost rang his friends Robert Stigwood, the impresario, and Paul Hamlyn, the publisher, and they took some shares. Michael Deakin had one friend he thought might be interested – Jonathan Aitken, a young Conservative MP, great-nephew of the press baron Lord Beaverbrook and chairman of a thrusting new investment company, Aitken Hume. Jonathan had worked for a while at Yorkshire Television, presenting current affairs programmes and documentaries. That was where he met Michael, a gifted film-maker and effervescent raconteur who was popular with most of those he worked with. They had become especially close in 1971 when Jonathan appeared at the Old Bailey accused of offences under the Official Secrets Act. The case concerned some documents about Britain's secret role in the Nigerian Civil War. Jonathan had used them as the basis for an article in the *Sunday Telegraph* but he had originally gained access to them for a Yorkshire TV programme that Michael was to have produced. Michael appeared as a witness in his defence. Jonathan was acquitted but it had been a worrying time and the two men formed a close bond of friendship and sympathy. A few years later when Deakin was charged with libel (and cleared) in connection with another programme, he instinctively turned for emotional support to Jonathan.

After leaving Eton and Oxford (he was a couple of years behind Peter at Christ Church), Jonathan established a political base by becoming private secretary to Selwyn Lloyd, the former Foreign Minister. Then he became a journalist with one of the Beaverbrook papers, the *Evening Standard*. He worked briefly on the management side at Beaverbrook, but his first foray into the world of finance came in 1973, when at the age of thirty he joined Slater Walker, the ill-fated investment company, as their managing director in the Middle East. Contacts he made in those days led to a major investment by the Saudi royal family in Aitken Hume. That firm was formed in 1981, partly from Beaverbrook money inherited by his cousin Timothy Aitken, Beaverbrook's grandson and now Jonathan's business partner. Coincidentally, their friendship had also been cemented by Jonathan's secrets trial. Timothy often sat in court and provided moral support.

The two Aitkens are a curious blend of personalities. The lanky, cultured Jonathan is smooth, transparently sincere, his hard centre scarcely detectable beneath the polished veneer. Few ambitious businessmen can flourish without the occasional sacrifice of a colleague, but if Jonathan must deliver the killing blow he does it much more in sorrow than anger, making plain that it hurts him more than it wounds the victim. If he has a choice, however, he prefers to leave the ruthless stuff to Timothy, a short, energetic tycoon of thirty-nine, who appears positively to revel in cut and thrust, affecting a curt and brisk manner that reminds people of his grandfather: indeed he often refers admiringly to Lord Beaverbrook, who was his stand-in parent after his father died in an accident in 1947. Those who have experienced the joint assault of the Aitkens tend to speak of them as the pair of interrogators from the best traditions of the spy novel, the one polite and regretful, the other bristling with threats. Jonathan's attitude to Tim is that of an owner with an irascible Alsatian: it is true that he devours the occasional small child, but he certainly keeps the burglars away.

The pair have a third partner, Michael Scorey, with a reputation as a financial wizard. In discussion of company strategy the trio decided they would concentrate on buying into media companies, especially newspapers. They felt it appropriate that money inherited from Beaverbrook should be put into such ventures, perhaps one day recreating the press empire that had been forfeited – largely, they believed, through prior mismanagement – when Lord Matthews bought the Express group in 1977. They would speak of their 'ancestral link' with the press and 'buying back the family farm'.

But there are not many national newspaper groups around – and the redoubtable Rupert Murdoch owns the most important of them – so the Aitkens had decided first to have a try at television. The IBA's invitation to applicants for a breakfast proposal was appended to their regular eight-yearly review of all the local independent television franchises. The Aitkens had become involved in two of these regional bids, most importantly in Yorkshire where, in a consortium that included former Premier Sir Harold Wilson, they were trying to depose Yorkshire TV.

So when Michael Deakin lunched with Jonathan and asked whether he would be interested in investing in TV-am, it did not take him long to decide that yes, he would be. Jonathan phoned Dick Marsh, and soon met Peter Jay and David Frost. The Aitkens were willing to put in up to £2 million, which would have given them about 25 per cent of the company. But Peter, David, Dick and Michael Rosenberg had already decided that no single investor should be allowed more than 15 per cent and they kept them to that. One of the features of their plan was the element of ownership by the founder-presenters. They could not realistically be major shareholders, but at least they wanted to ensure that no other person or institution should find it easy to assume too firm a degree of control.

So the Aitkens made do with 15 per cent. The restriction did not bother them too greatly because, competing against seven other groups, some backed by names as distinguished as those involved in TV-am, there seemed only an outside chance that the franchise would be won. Jonathan had rather more confidence in his bid in Yorkshire.

TV-am's application document was a joint production. Michael Rosenberg wrote most of the financial part, Dick

Marsh the piece about labour relations and a Canadian named
Tom Cook, hired as a consultant in the consortium's early stages,
took care of the technical specifications. Nick Elliott composed
a sample programme schedule. It was left to Peter to write the
part likely to weigh heaviest with the IBA, concerning the
programme content. 'Their governing philosophy,' he wrote
of the gang, 'has been to create the capacity to produce a new
form of television; and all of their actions, including the
recruitment of the ablest and most experienced programme-
makers, have sprung from this. . . . All five members of TV-
am's on-air nucleus combine a proven journalistic background
with an established popular following.' As for Peter himself,
he was 'exceptionally qualified to oversee the translation of
TV-am's ideas and ideals into a practical and successful service
to the public.'

These ideals, according to the prospectus, were to interest,
inform and divert in equal proportions and to 'open a dialogue
with their viewers, playing an increasing role in enriching the
quality of their lives.' And the United Artists concept of
management was specifically highlighted. 'Broadcasting is the
proper concern of broadcasters; and the editorial and financial
control should be in the hands of those qualified in both respects
to make the professional judgements necessary to mount the
most exciting challenge of television of this decade. TV-am feel
that the spirit of a co-operative of highly experienced and
responsible broadcasting professionals, both in front of and
behind the camera, is the proper inspiration for British television
in launching this new development.'

The confident tone of those lines did not entirely reflect the
mood of some of the consortium members, none too ready to
burn their boats with the conventional TV stations where they
had made their reputations. Of the presenters, neither Esther
Rantzen nor Angela Rippon would do anything so rash as to
announce their involvement in this risky venture until they knew
for certain that the franchise would be won. Michael Deakin
and Nick Elliott also opted for anonymity at that stage, and
so did Anna Ford when she joined later that year.

The IBA takes itself seriously as a guardian of the public
interest, and before awarding franchises it undertakes a

13

process of consultation in the form of public meetings where viewers can come and quiz applicants for franchises about their intentions. This makes some sort of sense in the context of the regional stations but it was an unwieldy mechanism for regulating the national breakfast bid. To overcome the geographical handicap they organised two meetings specifically to discuss breakfast television, one in Croydon, south of London, and the other in Darlington, chosen because it was handily reachable on the main line from London to Edinburgh. It is hard to inspire public interest in such meetings and in both cases – but especially in Darlington – the representatives of the eight companies felt that they, combined with the press and other TV professionals, outnumbered the general public.

Little concrete was achieved, but the Croydon meeting was notable for an inspired piece of phrase-making from Peter. As a result of his persuasive advocacy at their strategy meetings, and because they felt it was the kind of thing that would go down well with the IBA, the gang had decided to make the Jay-Birt thesis central to their programme-making philosophy. They would be producing a revolutionary brand of current affairs television, a unique and judicious mixture of news and interpretation that would eliminate the bias against understanding that the two sages had identified five years earlier in their *Times* articles.

The trouble was that the phrase 'bias against understanding' is a ponderous one that strikes no immediate chord with the public. Moreover it is a negative description of the philosophy, referring to what it seeks to put right rather than to promote. (Indeed, Peter had more than once to correct press interviewers whose questions indicated that they thought he was in favour of incorporating a bias against understanding into the programmes.) At Croydon, while he was on his feet giving an impromptu account of the TV-am philosophy, the right words came to him. What he and his colleagues were embarking on, he declared, was a 'mission to explain'. Lest that should sound too weighty he insisted then, as he did whenever he wheeled out the phrase in future, that it would go in harness with an equally powerful mission to entertain. The comparison he liked to use was with the *Daily Mirror* of the 1950s, or 'Sydney

Jacobson's *Daily Mirror*, as he would call it. He held it up as an example of how to combine sound factual reporting and political analysis with popular entertaining features and package them for a mass market.

It was an unfortunate image for a number of reasons. To begin with, the 1950s was rather a long time ago. Moreover, Sydney Jacobson, now Lord Jacobson, was never editor of the *Daily Mirror*. For part of the 1950s he was its political editor, before moving to be editor of the *Daily Herald* and its successor the *Sun*. And this is where the comparison turns sour, for both the *Herald* and the original *Sun* were notably unsuccessful attempts to make responsible journalism popular. The *Sun* was eventually sold to Rupert Murdoch who turned it into a roaring success by abandoning serious news analysis and making it a less demanding read than the *Mirror*, as well as by introducing daily pictures of naked women. Faced with this competition the *Mirror* had to follow the *Sun* down market, quickly jettisoning its serious pretensions in features such as Mirrorscope and the Inside Page. What the episode proved was that serious popular journalism cannot compete successfully with unserious popular journalism – a lesson Peter was to learn more painfully when his station went on air.

For the time being, however, the mission to explain was to stand Peter in good stead at the mandatory interview with the IBA. This took place on 11 December 1980, just over two weeks before the decision was due to be announced, at the Authority's headquarters in Knightsbridge. Peter was in his best magisterial form. Describing breakfast TV as 'the next great frontier of television', he spoke of the new kind of TV journalism that was so urgently needed, to make the world and its happenings come alive in a way that the traditional news media had failed to do.

There was one important exchange at the interview which was to assume importance – and be subject to conflicting interpretations – later on. Lord Thomson, formerly Labour Cabinet Minister George Thomson, was vice-chairman of the Authority, serving his apprenticeship before taking over from Lady Plowden as chairman the following year. He quizzed Peter about the structure of the company. At that stage Peter was part-time chairman. The group's intention was that, if they won the franchise, a chief executive would soon be appointed. Most of the other contending consortia had a chief executive on their team already. Thomson saw a danger signal here and asked a blunt question: 'You have an extremely brilliant and high-powered executive chairman, so once you have a chief executive will you not have a bit of a problem about keeping out of his hair?' Peter, having patronisingly told Thomson that his question was 'an extremely proper one', went on to articulate the Panglossian optimism that is a hallmark of his thinking: 'I believe these things in human affairs work out extremely well given that you choose the right person who complements your skills.'

Historically, this was a vital exchange. For when, as the months rolled by, no managing director was appointed (partly to avoid stretching the start-up budget), Peter's supporters were able to refer back to the interview as evidence that the IBA thought they should not appoint a power-sharing chief executive at all. Thomson was later adamant that he had not meant that, and the record seems to bear him out. In any event, the failure to appoint a managing director was a crucial element in the dispute that erupted two years later. But the most prophetic observation in that interview was made by Angela Rippon. 'If you give us the franchise and we do not succeed our reputations are at stake.'

The interview had gone so well that Peter, who had in any case always been confident that he knew precisely how the franchise could be won, was now increasingly convinced that they were going to win it. At their late lunch in the Hyde Park Hotel following the interview, his ebullience infected the others. A few days afterwards he was invited to dinner by the iconoclastic former Labour MP Woodrow Wyatt at his house near Lord's cricket ground. Among the guests were Harold Lever – chairman of one of the rival groups bidding for the franchise – and Lord Weinstock, chairman of General Electric and one of Britain's most highly-regarded industrialists. Peter remarked to Weinstock that he had the feeling he might soon find himself in charge of a small company and would welcome advice. 'Are you going to be chairman or managing director?'

Weinstock asked. 'Suppose for the sake of argument I'm chairman,' Peter replied. 'Then appoint a finance director to spy on the managing director,' snapped the industrialist. 'And if I'm the managing director as well?' Peter inquired. 'Seize the petty cash and allow nobody to sign a cheque.' Peter did in fact stick to that second piece of advice for a considerable time, for he felt it vital to inject a strong sense of financial discipline into his company, knowing that in the world of the media costs can often get out of hand. When Jonathan learned about the dinner table advice, he would tease Peter about it remorselessly.

David Frost had arranged a lunch party at his Kensington house on the day the IBA's decision was due to be announced, the December Sunday in that normally dormant week between Christmas and New Year. Peter went to the Knightsbridge headquarters with the leaders of the other seven consortia to collect their sealed letters from Lady Plowden. Only one would contain the magic key to the TV station and just after 2 p.m. Peter phoned the house with the news that it was theirs. There was jubilation, the cracking of fresh supplies of champagne, then an invasion by the press and television cameras.

In the excitement, nobody took much notice of the proviso in the contract award that the new station would not be allowed to start broadcasting until some time in 1983, instead of January 1982 as had been envisaged originally. It meant they would have to wait more than two years before the programmes could begin and – more vitally – before they would start receiving revenue. But the party atmosphere would not have been receptive to such quibbles, or to any expression of the growing reservations some members of the gang were having about their role in the enterprise. It had been fun while an insubstantial idea, one of eight proposals to the IBA and not the one predicted to win by the bookmakers. Now it had become a reality. It was actually going to happen. For some at David's house, it would be time for sober reassessment as soon as the party was over.

After that euphoric Sunday a board of directors was constituted and had their first meeting in David's dining room. Encouraged by Lord Thomson's ambiguous intervention during their interview, they decided to appoint Peter chairman and chief executive, with the intention of finding a new chief executive shortly before they went on air. Only Dick Marsh, made vice-chairman, wondered whether somebody with more business experience than Peter ought to share power with him, but his reservations were ignored in the heady atmosphere, where it seemed somehow disloyal to believe that the venture was going to be anything other than a brilliant success from the start. Soon a suite of prestige offices was found in Deanery Street, Mayfair, hard by the Dorchester Hotel.

It is almost certainly a mistake for two people with similar natures to go into business together. The Aitken partnership works because the pair complement each other with contrasting strengths and weaknesses. David and Peter share a clearly attractive but, in business terms, an ultimately fatal optimism. David would make no claim to equal Peter in intellect, while nobody would accuse Peter of possessing David's sharp wit and *bonhomie*. But they do share a belief that if people of good will are in charge, everything will magically come up roses – and they cling to that tenet long after the evidence starts to show that what they thought were roses are really thistles, and pretty intractable ones at that. Of the other board members, those who were not friends were representatives of the financial institutions that had invested in TV-am and for the most part this was their first experience of anything to do with the entertainment industry. If they thought they detected a lack of rigour and conventional business practice, they could shrug and tell themselves that this was, of course, show business and that must be how things are done. Even the Aitkens now had the wrong man on the board. It had been Timothy before the contract was awarded but afterwards the smooth, amenable Jonathan, more experienced in television, took over, with Timothy as his alternate director. Jonathan is not keen to make ripples until he feels there is truly no option. When Timothy did attend meetings he invariably managed to upset the friendly, confident mood and he was not liked for it – but that only happened once or twice in the first eighteen months. The sole sceptic on the team was Dick Marsh, but as he was a minority of one it was too easy for his jeremiads to be ignored. Peter and the others used to chuckle and call him 'Mr Any Questions,' implying that his cautionary speeches were not really meant to be acted upon, but were of the kind calculated to gain the approval of the reactionary burghers of country towns who generally form the audience of that radio discussion programme.

Thus when things began to go wrong in the early stages, the gang carried on regardless. The first blow was when Esther Rantzen became pregnant and decided that being the mother of a young child would not be compatible with getting up in the dark to go on air at 6.30 in the morning. She had in any case not resolved in her own mind that she really wanted to give up 'That's Life', the show she had shaped almost single-handed. So she pulled out – but that was all right because Anna Ford had already been hired. The Famous Six were back to their original strength of five. Then Nick Elliott won a long-awaited promotion at London Weekend and decided to accept it rather than stake his future on what he was beginning to see as a highly speculative venture. But that was all right, too, for there had been a bit of uneasiness over the division of responsibilities between Nick and Michael Deakin. Nick was called Director of Programmes and Michael Director of Features, and that menacingly ambiguous phrase *primus inter pares* had to be deployed to define their relationship. The departure of one of them made the chain of command that much more straightforward. Everything in the garden was still lovely, even if the garden was getting a bit depleted. Judicious pruning, after all, is known to strengthen the stock.

Another setback was the failure to reach final agreement with Ewart over the Wandsworth studio. Just whose fault that was is disputed. Ewart can be a disconcerting man to negotiate with but he remained insistent that he was ready to come to terms. Again, that was soon turned into good news when the gang found an old garage in Camden Town, backing on to Regent's Canal. Their architect, Terry Farrell, said the garage would be ideal for conversion into a brand new, custom-designed headquarters, complete with giant eggcups on the roof. They were like a newly-wed couple who, reluctantly resigned to living with their in-laws for the early years of their marriage, are suddenly presented with a place of their own and a large credit at Heal's.

Most of 1981 was occupied with finalising the contract with the IBA – a process delayed by a long dispute over whether the morning hours could be pre-empted by Independent Television News on big national occasions like general elections. Throughout this time Peter, keen to begin taking in revenue, was pressing the IBA for an earlier start date. He finally succeeding in having it advanced by a mere three months, from May 1983 to February. He was also negotiating with ITN over the provision of a news service. This was an important political question, for ITN had been among the applicants for the franchise but had been passed over by the Authority in the interest of diversity, of introducing a 'third force' in the news to complement the BBC and ITN. But the IBA were keen that whoever won the franchise should co-operate with ITN, particularly on foreign news, rather than set up an expensive news service from scratch. In their application TV-am had expressed willingness to do that but the negotiations proved tricky.

ITN were asking £6m for a complete news service. That was not what Peter wanted, even if he could afford it, since he was determined to integrate news reporting into his programmes. Part of the bias against understanding, he thought, was the existence of two separate commodities called news and commentary. By allowing ITN to slot ready-made news bulletins into his programmes he would be accentuating that division. He wanted 'raw' news from ITN which could be reported and commented upon by TV-am at will. An earnest debate began in which the phrases 'sovereignty over inputs' and 'sovereignty over outputs' were bandied about. Peter was in his dialectical element, but his mission to explain his viewpoint to ITN was unsuccessful. It was scarcely a surprise, when you thought about it, that ITN were in no mood to be accommodating, not just because their pride had been dented by the rejection of their own application but because Peter's high-toned strictures against existing TV news coverage had been aimed largely at them. The negotiations were stalled for several months. Only towards the end of 1982 was an acceptable accommodation reached. Peter said it amounted to the level of co-operation he had always wanted but others – including some of the professional staff at the IBA – thought it would place too heavy and expensive a news-gathering burden on the fledgling station.

They moved offices from Mayfair to a block just opposite their new Camden Town site, so they could watch the studio take shape. Key staff were recruited on the programme and commercial side. Many of the programme people came from Yorkshire TV, Michael Deakin's former company, forming the nucleus of what the junior staff were later to dub 'the Yorkshire Mafia'. A young Yorkshireman, Hilary Lawson, still only in his twenties, was hired as editor-in-chief. Initially he was to head the features side of the operation with Bob Hunter, an Ulsterman from ITN, as his opposite number for news. But Hunter had personality clashes with Michael and resigned before the station went on air.

There was a miscalculation when it came to recruiting the junior staff. Saatchi and Saatchi, TV-am's advertising agents, designed a full-page advertisement that appeared in the 'Creative and Media' jobs section of the *Guardian*. (Some thought that the choice of that newspaper, alone among the national dailies, was peculiar for a station that sought to emulate the *Daily Mirror*.) The advertisement contained full-length pictures of the Famous Five and Peter, surrounded by people-shaped spaces waiting to be filled in. The headline was: 'Work alongside us and make history' and the text invited applications from 'the best and the brightest young television journalists' as well as experienced technical and operational staff. So effective was it that thousands of hopefuls applied, far more than had been anticipated. All were asked to complete long, complex application forms. The machinery was scarcely able to cope. Many applications were not even acknowledged and the interviewing took much longer than had been allowed for.

When the final appointments were made it was apparent that there was a gap on both the programme and the technical sides between executives and the lowest-rung personnel. Middle-rank people were not coming forward, because the ITV companies in London paid more than TV-am were offering and the BBC was aggressively combating attempts to poach its staff by giving pay increases and promotion to many who said they were thinking of leaving. By now the BBC had announced that it was planning its own breakfast TV service, to go on air two weeks before TV-am, and the BBC executives, in their characteristic pugnacious way, were determined to wage competition fiercely, without conceding advantages to the enemy. Several BBC people whom Michael thought he had secured could not in the end bring themselves to sacrifice the comfort of the Corporation, any more than Esther Rantzen had been able to. So those hired were for the most part short of television experience. Peter, David and Michael managed to turn even this disability into a virtue by telling themselves that since they were committed to making a new kind of television it was a positive advantage to have people whose preconceptions were not anchored rigidly to the old ways. They were convinced there was nothing exceptionally bright people could not turn their hands to with success. Realistically, Michael told himself that even if up to half fell by the wayside, the rest would form the nucleus of a glittering team he could mould in the company's image.

Meanwhile the presenters were travelling the country promoting the new station to advertisers and the press. It was during one of these sessions that David invented − or had put into his mouth − the phrase 'sexual chemistry' to describe the mood the presenters would try to create on screen. In off-screen reality, there was genuine disharmony within the senior ranks. At root it was a question of management style. Peter, partly no doubt because of his Treasury and ambassadorial experience, believed strongly in written communication between colleagues. He liked to make sure that everyone was properly informed of decisions taken, of who had taken them and the reasons for them. Michael was the precise opposite. His ideal organisational structure was a group of people sitting around on the floor bouncing ideas off each other. Thus when Peter asked him to prepare a written outline of the programmes, Michael would reply: 'It's people who make programmes and we haven't hired the people yet.' But even when the people were hired, no such outline was forthcoming. Essentially, Michael did not take the mission to explain with the seriousness Peter did. He saw its attraction in preparing the application, but as for its practical value in making programmes . . . well, it was back to people again. It was all in the fingertips. Rigid formulae were anathema.

Michael saw life as a theatre and appeared to think of himself as both critic and leading player. He watched events unfold with a kind of knowing amusement, relating them to his global philosophy of the nature of institutions, individuals and power, particularly power. As an actor in the drama he could scarcely remain wholly detached from its quality and success, but he could protect himself by blaming any deficiencies on the author or the stage management.

Thus when Peter wrote a series of fifteen administrative memoranda, numbered sequentially, about such procedures as laying out letters and sending minutes up to the board, Michael waved them aside, calling them derisively 'the paper mountain'. His ironic sense of humour made him a popular companion and he would raise lots of laughs with stories of Peter's precise attention to such details, presenting it as a real-life parody of Treasury bureaucracy. (Coincidentally, some of Peter's clashes with John Robinson, his deputy at the Washington embassy, had been over Robinson's reluctance to compose long memoranda.) Michael treated Peter as some kind of pet extra-terrestrial who must must be humoured but not taken seriously. He would address him as 'Imperial Chairman'. Peter endured it all stoically. Whatever else a public school education may do for you it certainly teaches you to join in jokes against yourself in a sporting and gracious spirit. So Peter would grin and bear it, consoling himself with the thought that behind this merciless teasing Michael kept a deep respect for him as the boss and did not really mean it.

It was not only Michael who crossed swords with Peter. The most absurd but still the most heated of the pre-start disputes arose because Peter was determined that the design of the studios should incorporate what he called transparency. Again it was essentially a matter of communication. He believed strongly that organisations worked best where everyone knew what everyone else was doing and why. He spoke with horror of those factories where people made components for machinery without understanding what the components were for. That was, he felt, one reason for the malaise of British industry, and it would not happen at TV-am. In most TV stations, in his experience, the studios are rigidly separated from the offices and people working in non-programme departments are never privy to the mysteries of what goes on within them, until they see the finished programmes. So he insisted on two major features in the studio design: one was a bridge above the studios with glass walls so that people could see into them; the second was that glass walls be built between the studios and the control rooms. Michael and the technical people opposed both these ideas, especially the viewing bridge. They said it would hamper operations. Peter heard that the idea had worked at the ITV station in the Channel Islands, and the IBA arranged for him to go there and see for himself. He came back more determined than ever, and the windows were built; but because of scenery and other studio clutter not much can now be seen from them.

As the new staff arrived in weekly waves, Peter called them together for a welcoming address, encapsulating his philosophy of human relations in industry. Everyone, he said, had a real role to play 'because you are doing it with friends, all of whom are as nice and as talented as you are yourselves.' He pressed home the theme. 'An identikit TV-am person is (a) extremely talented and (b) extremely nice.' As for the mission to explain, Peter said they could all take copies of his lecture 'What is News?' which he had given at the Royal Society of Arts the previous March. It was a weighty document that began with the formidable phrase: 'Anyone trained in what was called "philosophy" at Oxford in the late 1950s. . . .' Those talented and nice new employees who struggled beyond that opening will have found little practical guidance on what was expected of them, save for a cerebral section on the reporting of traffic jams, suggesting that 'the density of traffic on any given route tends to vary inversely as it was the same time the day before.'

While his nice reporters were rushing off to test this novel theory on the traffic system that swooshes round Camden Town, Peter was having trouble with his board, who for the first time were showing themselves to be less nice than he had thought. It began in the summer of 1982 when Timothy Aitken, deputising temporarily for Jonathan, received the clear impression that several departmental heads had serious doubts about whether they would be able to keep within their budgets while still doing the job expected of them. Peter, following Lord Weinstock's dinner-table advice, had been enforcing strict budgetary limits and was so stern about it that the doubters had felt intimidated about raising their worries with him. Timothy encouraged them to confide in him while Peter was on holiday in August and Jonathan in hospital after cracking five ribs in a fall in the bathroom of a New York hotel. Returning tanned from his month in Sardinia, Peter found himself the target of a series of unexpectedly searching questions at the monthly board meeting. He had no answers ready but promised to undertake a thorough review before the next meeting. There he produced a frankly self-critical report in which he confessed that the budget figures he had given to the board earlier in the summer were defective and would have to be revised. He wrote of 'gross errors' in the May figures, of 'a failure of management at the highest level,' and accepted full responsibility for it. It was, he conceded, 'a grave position to report to any board'.

Some directors saw this document as a correct and honourable bid by Peter to take the whole blame for accidents that could not be entirely his fault. It proved, they thought, his strength of character and sense of what was proper. For the Aitkens, however, it confirmed their hardening impression that Peter was not the right man for this job. Doubts had been growing over the last year or more. There was the early collapse of the arrangement with Ewart, and the failure to hire several promising people whose appointment had been forecast to the board. And Jonathan was growing increasingly impatient with Peter's long speeches at board meetings. These were almost inevitable because with few executives on the board he had to make the reports that in other companies would be made by the finance director, the director of operations and others. Peter is a precise but discursive public speaker and his speeches went on too long for some board members. (During the 1980 battle for the franchise Ned Sherrin, one of a rival group, would joke that if Peter was allowed to do any interviewing, you could time a four-minute egg by his opening question.) Jonathan once observed to Michael Scorey, the third partner in his businesses, that Peter was like a man who speaks a language — in this case the language of business — absolutely fluently and beautifully; yet after a time you suspect he has not truly got the hang of the grammar.

With Peter's 'confession' as powerful support for their case, the Aitkens were able to convince the board that the time had come to remove him as chief executive, appoint him non-executive chairman and hire someone with proper business experience. Peter opposed this vigorously, saying all he needed really was a finance director. The Aitkens persuaded the board that it had to be a chief executive and they should start looking for one without delay.

On the weekend after the board meeting David Frost, worried that if Peter's authority was too greatly diluted he might pull out altogether, acted as a broker between him and Jonathan to work out a formula that would leave Peter clearly in command. In this he was bolstered by an assurance from Lord Thomson, chairman of the IBA, that the Authority regarded Peter as an integral member of the group that had won the franchise and would be worried by any move to depose him. After some days of negotiation, Jonathan worked out an acceptable formula that left Peter in charge of the company in its relations with the outside world while the new person – not now called chief executive but managing director – would run the company day-to-day. A firm of executive head-hunters was appointed. A number of candidates were interviewed but the quest took longer than expected. With the on-air date approaching, Jonathan took the initiative in suggesting that the new appointment should be postponed until the launch was completed.

The BBC's rival show, 'Breakfast Time', began in mid-January, two weeks before TV-am's start date. It was surprising in two ways. First it was a brazenly down-market show, placing more emphasis on breezy features than on hard news reporting. The two presenters were Frank Bough, whose rugged, cosy face had become familiar through hosting BBC's Saturday afternoon sports coverage, and Selina Scott, an attractive young woman who had replaced Anna Ford as an ITN newscaster. And they were the second surprise, for they settled down straight away into an easy, comfortable relationship. Some eight million people were estimated to have watched at some time during that first morning and many liked what they saw.

When it quickly became clear that the BBC were going to put on a rival breakfast show – despite constant pleas that they needed a higher licence fee to fund even the shows they had before – everyone assumed that they would take their cue from the declared intentions of TV-am and launch a mission to explain things even more clearly than Peter's people were promising. It would, they thought, be a morning version of 'Newsnight,' the sober BBC 2 late-night news analysis programme. But when they saw they were competing against the *Daily Express* rather than the *Daily Telegraph* they had expected, the people at TV-am remained confident that they could outclass them. How could they fail, with stars like Anna, Angela, David, Michael and Robert? Surely it would be no contest? And even when their two weeks of 'dry runs' revealed serious flaws both on the technical and programming sides, their confidence was scarcely dented.

The first programme, on 1 February 1983, was better than the run-throughs. Indeed, it was not at all bad, in spite of being seriously imbalanced by David Frost's seventeen-minute interview with Norman Tebbit, the Employment Secretary. (The interview should have been less than half that length.) There were some good ideas – notably a merry feature called 'Through the Keyhole' in which an American journalist made offensive comments about the homes of famous people without being told their identity. And there were some rotten patches, especially a camp five-minute soap opera called 'The World of Melanie Parker', which misfired dismally yet was not scrapped for six weeks. As for the mission to explain, nothing was detectable that warranted any such description.

Judging the success or failure of a TV programme is a delayed-action process. It takes a week for the ratings to be worked out and published. For the first few days, then, the TV-am team scarcely realised that they had been anything but the runaway success they had hoped. The first week showed them with a peak-time rating of 800,000 – just half that achieved by the BBC. Ratings are a subject of endless controversy in the TV industry and in particular ratings for breakfast shows that may be watched for only a few minutes at a time. The rating system specially devised for breakfast television measured the audience in the peak quarter hour. While this is an inadequate guide to total viewing figures it does at least give some kind of yard-stick by which to compare the two services.

The Famous Five and Peter had together, many months earlier, worked out a complex roster for presenting the main programme, 'Good Morning Britain', that followed the hour-long news-based section called 'Daybreak'. The Five would work in pairs. All of them were on deck for the grand opening but the main presenters of 'Good Morning Britain' for the first month were to be David and Anna – David because he was the father of the whole project and Anna because she was thought to be the most appealing of the other four.

As ratings dropped to 500,000 in the second and third weeks, then to 300,000 in the fourth, nobody dared suggest that the personalities of the presenters, the basis of the entire enterprise, might be a contributory factor. Other explanations were sought. The ratings were said to be defective. They did not, it transpired, take account of TV sets in hotel rooms where, the argument ran, businessmen in their thousands were switching to TV-am every morning, quite unrecorded. Then the advertisements were said to be the trouble. People were apparently turning over to the BBC as soon as the first commercials appeared. If that was so they could scarcely be blamed because the ads were unusually dismal, mainly because of a long-running industrial dispute that affected TV-am more than the regional ITV stations. It was between Equity, the actors' union, and the Institute of Practitioners in Advertising, who were trying to get actors to accept lower royalties for advertisements appearing on both TV-am and the new fourth channel, because of the low viewing figures. As a result of the dispute few commercials featuring professional actors appeared on the screen. Instead, we had a series of charmless businessmen extolling the virtues of their products. One that irritated by constant repetition was for Wall's sausages and bacon: at the time of writing it is still sometimes to be seen.

As the audiences continued to sink, some did begin to question the performance of the presenters. Michael Deakin, for one, thought it increasingly clear that David and Anna did not work as a team, that any chemistry was producing dross rather than the alchemist's gold. His analysis was that Anna, with hardly any experience in unscripted television, was naturally nervous and, hard as she might try, came across as cold and withdrawn. David, who likes nothing more than making it up as he goes along, sensed a wrong atmosphere and overcompensated, going over the top in heartiness. He made one well-publicised crack about the water workers' strike then under way, inviting viewers to boil water and throw it over the strikers. The Valentine's Day edition was acutely embarrassing, with Anna and David reading silly valentines culled from the personal columns of the daily papers. Anna read out one to 'wobble-bottom' and, told she must try and be more jokey with David, added: 'Could apply to you, couldn't it?' David, with what can only be described as a snigger, replied: 'What a vile slur — and what an implication of sexual chemistry.' As someone who worked on the show put it: 'It seemed as if Anna was driving David crazy' — not at all the kind of chemistry that had been intended.

The presenters, on the other hand, blamed the programme staff for the low ratings. The heaviest weight of their anger fell on the shoulders of Hilary Lawson, editor-in-chief. Again, there was something in that. They had been given some fairly dismal items to present. Two had particularly embarrassed Anna — one about female circumcision and another an interview with an American who had written a book about the G-spot, allegedly the most sensitive part of a woman's anatomy. The items seemed to be flung together with little thought of where in the schedule they might be appropriate. Thus in that Valentine's Day special, David had just read yet another personal message, this one from 'snorty-bloggs', when it fell to Anna to introduce an item about rape. 'In London alone a woman is raped or violently attacked every hour,' she had to intone, in a grating switch of mood. The programme ended with the slow movement of Beethoven's Romeo and Juliet.

Anna did her best but failed to overcome the clear impression that her idea of the programme was Beethoven rather than snorty-bloggs. An unmistakable air of condescension emanated from her, try as she would to conceal it. Dick Marsh had spotted this and, asked by a reporter to comment on the early failure, remarked with wicked accuracy that it was like members of the Royal Shakespeare Company presenting 'Listen with Mother'.

The presenters, especially Anna, were furious, and made certain that Peter knew it.

At the end of the first month, according to Peter's master roster, it was the turn of Anna and Angela to host the show. They promoted this all-woman team heavily on the ITV network, although some of the staff had serious reservations about whether that was an ideal combination, either. There was scant time to test it for, on the evening after their first show together, Anna phoned Michael Deakin to say she had 'flu and would not be in the next day.

David stood in on her first day's absence and Michael Parkinson for the rest of the week. With Angela and Michael the programme looked better than it had ever done. Michael was already getting decent ratings for his weekend shows – not subject to competition from the BBC – and he and Angela worked together smoothly, old troupers both. Derek Stevenson, in charge of advertising sales, was especially impressed with the new pairing for he had been getting depressingly negative responses from advertisers to the previous combinations and in particular to David. So he went to see Michael Deakin to discuss whether there was any chance of changing the roster to keep the pair together. Michael was keen, but it would have meant resting Anna, due to return from her illness the following Monday. After discussing it with Peter, Deakin rang Anna to put the proposition to her. She thought about it for a while, then phoned back and said she was determined not to stand down. Peter went to her house in Brentford that evening but could not get her to change her mind.

The incident depressed Deakin. Here he was, nominally the director of programmes but, because of the Byzantine commercial structure, unable to get his way on the basic matter of which presenters should go on air on which days. The presenters were founder shareholders, paid more than he was. They had worked out their schedules with Peter, so only Peter could change them – and he wouldn't. Part of the problem was that the presenters had been led to expect much more authority than they could realistically wield. And, ironically, they had been encouraged in that by the reference to United Artists in the IBA application – which had been Deakin's idea in the first place. They were exercising what would come to be known as 'presenter power': and power was something Deakin set great store by. While working for Yorkshire he had been based in London, and used to liken himself to the Duke of Burgundy, working out his relationship with Paris (head office in Leeds) while retaining maximum autonomy. Questions of power had been behind his difficulties with Nick Elliott earlier on. True, life was a theatre. But it was one in which he was determined to remain at centre stage.

Deakin also had a sardonic streak. An example just a year earlier was a caustic review he wrote in the *Spectator* of the autobiography of Alan Whicker, the veteran TV interviewer with whom he had worked at Yorkshire. This was a sustained and mocking piece of invective and it included three sentences that some would see as bearing relevance to the current dispute: 'In the end, people always ask, why do such egos, such unlovable stars, survive in show business? The answer is that the system supports them. Directors and researchers are used to receiving no credit and less thanks (it's part of the job).'

The row over Anna occurred on a Friday. On the Sunday Deakin sat and wrote a plaintive letter to his old friend on the board, Jonathan Aitken. Later to become known as the 'smoking gun letter,' it was a scarcely disguised appeal for help. He described the Anna Ford incident and maintained she had threatened to resign if she did not appear on Monday and that Robert Kee would have supported her. If the ratings should fall he expected an assault from some of the presenters on the programme. He anticipated choppy waters.

Choppy waters indeed. The letter was all Jonathan needed to reinforce his growing conviction that Peter really must go. All last autumn's confusion about the budget and the other reservations he had about Peter's managerial style would have been insignificant had the product been a roaring success. It was not. It was a resounding failure. The ratings were so poor that advance advertising bookings were slowing to a trickle. The station had budgeted for a loss in its first year but was now faced, unless ratings improved, with a loss of such astounding proportions that no bank would continue to finance them. And now came this evidence – at least Jonathan saw it as such –

that Peter was incapable of controlling his staff, as well as being a faulty manager. He was now convinced that Peter's departure was essential not merely for the health of the company but for its very survival.

He called a meeting at his house in Lord North Street, just off Smith Square, of directors representing the institutional investors. Enough of them agreed with him to make him confident he would have a majority at a board meeting for dismissing Peter. The meeting was called for Friday, 11 March. Much of it was taken up with a long speech from Peter defending his record and promising that things would get better. It was a powerful speech and helped sway some waverers into supporting Peter. Some, too, had changed their minds after consulting their superiors. At the first meeting, for instance, Jonathan had the impression that Charles Wilson of Rothschild Investment Trust and Ben Martin of Barclay's Merchant Bank were on his side. But their chairmen – Jacob Rothschild and Lord Camoys respectively – decided they should keep Peter in place. The meeting ended without a decision. But Timothy and Jonathan, now they had started the business, were determined it should not end there.

Peter, too, knew something had to be done. He was now committed to the presenters' view that the fault lay not with them but with their backup from producers and editors. Earlier that week he had phoned Mike Townson, editor of Thames Television's weekly current affairs programme 'TV Eye'. Townson had been involved in one of the unsuccessful breakfast consortia and he had also applied for the job of managing director when TV-am had been planning to appoint one the previous autumn. He is a former *Daily Mirror* man who has been more successful than most in adapting the techniques of popular journalism to television. Brusque and autocratic, he did not have Michael Deakin's finesse and it was hard to see how they could work together. But since he had learned of Michael's letter to Jonathan, Peter felt no obligation to protect him. The needs of the station must come first. Peter met Townson for a drink at the Garrick Club on Friday evening, after the board meeting. Townson was interested in his proposition, though still keen to be made managing director

– not really the job Peter wanted him for. All the same, he had the impression that agreement was possible.

The following Monday Peter summoned Deakin to tell him of his decision. The two had already, the previous week, had a showdown about Michael's letter. Michael had insisted that the letter was not intended to undermine Peter. He had, after all, stressed in it that he had ultimately concurred with the decision over Anna. He had written it to defend himself against assaults from the presenters that he was convinced were coming. Peter pointed out to him that any outsider reading that letter would view it as a description of a wrong decision weakly made by a chief executive under pressure from a presenter.

They had ended that meeting on reasonably amicable terms but there was nothing amicable about Michael's demeanour when Peter broke the news about Townson. He went white, screwed up his face and snarled that it had now become a struggle between the two of them. He was not going to be the loser.

That was how things stood on the evening of Monday, 14 March, when, coincidentally, Peter had a long-standing invitation to dine with the Aitkens at their Lord North Street house. Jonathan's Swiss wife Lolicia has many friends from the European nobility and asked a group of them, visiting London, for dinner. Most were women – countesses and baronesses – so Peter, separated from his wife, seemed an ideal spare man to balance the numbers. On the morning of the party, recognising that Jonathan was scarcely ready to abandon the campaign for his removal, Peter telephoned Lolicia and, outlining the situation, asked if in the circumstances it would be better if he did not come. Lolicia was horrified at the thought, especially since his absence would have meant thirteen at table. 'Don't be ridiculous,' she told him. 'You can have your business fights but jolly well come to my dinner party.' Peter, anxious not to offend, rather reluctantly donned his dinner jacket and went.

Jonathan, greeting him, smiled and said: 'We won't say a

single word about TV-am.' It was a vain aspiration, for after
dinner they abandoned the baronesses and the other guests and
went off into the corner for a longish chat. After a while Lolicia
went to separate them. When they protested that they had
important things to discuss she declared with a laugh: 'Oh Peter,
Jonathan says you ought to get rid of half the presenters, but
isn't it true that you're really in love with Anna Ford?' Peter,
sporting as ever, grinned at this joke against himself and denied
it, but then he launched into a touching tale. It was true, he
said, that he had been greatly smitten with Anna when he had
first met her at Edinburgh. It was natural that they should see
each other quite often after that as the gang drew closer together.
Though wary of women after the trauma of his break with
Margaret, he felt a growing attraction but was not sure whether
Anna reciprocated it. Believing it important to end the
uncertainty he determined to face the question head on. He
would invite Anna for lunch and ask her how she felt about
entering into a relationship. The first date they could both
manage was ten days away, and the lunch reservation was duly
made. He was keenly anticipating the event. Devastatingly, on
the day before they were to meet, Anna phoned and said she
would be unable to make it. The reason, she announced
excitedly, was that she had just accepted a proposal of marriage
from Mark Boxer, the cartoonist. 'Congratulations,' said Peter,
and cancelled the table.

This tale affected the guests in different ways. Those
baronesses who heard were clearly baffled by this peculiarly
English love story, but Jocelyn Stevens, the magazine publisher,
laughed without restraint. Peter and Jonathan, their talk
interrupted, decided to take a stroll outside before Peter went
home. It was then that Peter told Jonathan for the first time
of his plan to appoint Mike Townson to strengthen the
programme side. Jonathan thought that was fine as far as it went,
even if it was going to ruffle the feathers of his friend Deakin,
but he was more concerned about whether Peter could deal with
the presenters. There was a school of thought – and Jonathan
agreed with it – that believed David, not Anna, to be the real
reason for the February disaster. He was not a morning person.
He looked drained on screen, with prominent bags under his

eyes. Someone had described him as being like a night-club comedian dropping into the studio and using up the jokes he had not got round to firing off during his night-club act.

Derek Stevenson had been getting this message from advertisers and pressed it on Jonathan, who was now convinced that the true test of Peter's mettle – his commercial virility if you like – was whether he could tell David to stay off the air for a few months. For Peter it was a painful prospect. David was, after all, the founder of TV-am and had recruited Peter to it in the first place. Yet although an entrepreneur he was still essentially a performer, with a performer's acute sensitivity about his own performance. In any form of show business, current affairs television just as much as the stage, the only acceptable form of criticism is: 'You were wonderful, darling.' People who cannot bring themselves to say that are expected to stay silent. For a friend to say: 'I'm afraid it's not quite working out' would be regarded as an act of hostility and betrayal. David and Peter were the best of friends and the cruel put-down is far from Peter's nature.

So when Jonathan put this question to him as they paced the square, Peter was able to say only that it should be left to him, that he would do what he thought right when he believed it opportune. David had been a loyal friend and colleague, he explained, and the harder Jonathan and Timothy sought to press him the more difficult it would be to take action. In any event, Peter added, he did not believe David was the chief problem. He had now become persuaded by the presenters that it was from the programme staff that the deficiencies stemmed. 'And I can't do anything about them because you're protecting them,' he complained.

Jonathan brushed all that aside. Whatever was the truth, he saw Peter's dithering over David as yet another example of procrastination, using fine words to delay doing what was necessary. Even at that stage, if Peter had given any sign that he was going to take a decisive grip, Jonathan might have persuaded Timothy that they should call off the move to oust him. But he felt that Peter was constitutionally unable to escape from his loyalties and friendships and commitments to other people. Jonathan thought of TV-am as an aircraft heading for the ground at an accelerating speed. It needed a pilot of great strength and nerve to pull it out of its nosedive and avoid a crash. Certainly if he thought the dive had been caused by the incompetence of his co-pilot, he should not worry about ruffling his colleague's feelings before seizing the controls himself.

And there was another thing. Peter's coolness under fire was beginning to irritate Jonathan immensely. There he was, being told in unmistakable terms that Jonathan was plotting to have him thrown out, and he was discussing it as calmly as if they were talking about house prices, the weather or any other conventional dinner-time topic. Jonathan would have had a great deal more respect for him if he had aimed a punch at his nose, right there in Smith Square. Suddenly he was reminded of a passage from *Heartburn*, that novel written by Nora Ephron, wife of the man Margaret Jay had her affair with. Ephron had written of the character based on Peter that he 'never takes anything personally; he always sees himself as a statistical reflection of a larger trend in society.' That, Jonathan thought, was Peter to a T.

They strolled for the best part of an hour. Then Jonathan walked home to Lord North Street while Peter drove to his larger but less fashionable residence in distant Ealing.

CHAPTER 2

I Know, Let's Start a TV Station

The conception of breakfast TV in Britain can be pinpointed to a precise time and place – the end of April 1979, in a poky office above a furniture store in Tottenham Court Road, London. Three television professionals were sitting round with not very much to do – a condition that in an articulate business generally leads to talk on the lines of 'Whither TV?'

Jonathan Dimbleby, David Elstein and Martin Smith had just returned from a long stint abroad filming 'Jonathan Dimbleby in South America' for Thames Television. All three had worked for some years on 'This Week', the Thames current affairs programme, Dimbleby as a reporter and presenter and the other two in production. But 'This Week' was about to be wound up. Jeremy Isaacs, its talented editor, had already left Thames. There had been criticism that the programme often took too committed a line on political and overseas issues. It was to be replaced by 'TV Eye' under the direction of Mike Townson, expected to concentrate on hard reporting rather than controversial interpretation and zealous campaigning.

Dimbleby is himself a figure who attracts controversy. He is the younger son of the late Richard Dimbleby, the plummy-voiced pioneer of current affairs television for the BBC, famous for his commentaries on royal occasions and for being the original host of 'Panorama', the prestigious weekly programme. Jonathan's elder brother David seemed to have inherited his father's mantle – and for a while his spot on 'Panorama'. He was the safe, polished professional, while Jonathan was of a more radical, uneasy stamp. On 'This Week' he had often been

37

criticised for supporting left-wing causes. Those Conservative MPs who from time to time accused television of being run by 'pinkoes' had such as Jonathan very much in mind. Reports by him and others from Ireland, the Middle East and Southeast Asia were felt in the industry to be one of the reasons for the demise of 'This Week'. In 1980 all the commercial companies were going to have to present themselves to the IBA for re-selection – an agonising and uncertain process that, mercifully, happens only once every eight years. The IBA is an appointed body, subject to political influence. Although Thames could not admit that the ending of 'This Week' was partly motivated by a desire to clean up their act for the re-selection lottery, that was inevitably how people interpreted it, especially those who had worked to shape the programme.

All this was running through the minds of the three men as they lounged in the West End office. As a preliminary to re-awarding the ITV contracts, the IBA had asked for suggestions about new franchises they might consider putting out to tender. The fourth channel was already in hand and due to begin broadcasting in 1982. Was there any other innovation to be tried?

Dimbleby, Elstein and Smith took up that point and wondered whether there might be an opportunity for a group to do their own thing, without having to depend on the patronage – and the political sensibilities – of a large company. In recent years there had been a lot of discussion whether news and current affairs television was done properly on either channel, and how it could be done better. The Jay-Birt articles were just one symptom of the general uneasiness in the industry on that point. Television in its present form had been developing for about 20 years but was it developing in the wrong way? The BBC was envied worldwide as an exemplary broadcasting authority that produced quantities of high-quality programming; but what they were most famous for was drama and historical serials, not news and current affairs. 'Panorama' was worthy but dull, and even the BBC's evening news was consistently – and deservedly – outscored in the ratings by its peppier rival ITN.

'This Week' and Granada's 'World in Action' had been regarded, at least until the mid-70s, as the most enterprising of the current affairs programmes on offer. The Jay-Birt articles were partly responsible for their losing some of their gloss, because the arguments were aimed chiefly at programmes like them. The first article, written by Birt alone in February 1975, used 'This Week' as an example of a programme that concentrated on a single aspect of a problem rather than the problem as a whole:

> Feature journalists tend to make a film about a particular instance of famine rather than about the world food problem. They expose the dangers of particular nuclear reactors rather than examine what the Government's energy policy is or could be . . . Making a film about homeless people is not an adequate way of approaching the problems created by our housing shortage . . . Television feature journalism continually suffers because producers do not take the trouble to think their ideas through and thereby to discover the tenuousness of the line between, say, one unemployed man and the real causes of unemployment.

Broadening the argument, Birt blamed such programmes for encouraging in the victims of social ills the belief that 'a sore easily highlighted should be a sore easily removed'. And he criticised studio discussion programmes that turned into gladiatorial contests between advocates of opposing viewpoints. To improve matters, Birt and Jay in succeeding articles drew up a blueprint for ideal network news coverage, with hour-long daily, weekly and monthly programmes putting the news in context. In one of them Jay invoked the image, which he has often used since, of a child's face dawning with understanding when something tricky was explained, saying: 'Oh, *now* I see.'

Dimbleby, Elstein and Smith could scarcely be expected to accept that particular recipe for doing their jobs better, given that their programme had been explicitly criticised. Like many, they thought the type of television suggested in the articles would, however worthy, be terribly dull. On a philosophical level they thought the notion that events could be reduced to a definitive explanation for handing down to the populace was dangerously didactic.

Birt and Jay, however, were not the only people questioning

the techniques of TV news and current affairs. A more directly political assault had come from the Media Group at Glasgow University in their book *Bad News*. They argued that the news on TV, far from being undermined by left-wingers, was in truth based on middle-class or Conservative assumptions. They asserted that industrial correspondents reporting strikes invariably let a tone of disapproval creep in. In interviews, reporters would hector strikers but be subservient to management representatives; they would concentrate their reports on the effects of strikes in terms of inconvenience to the general public, rather than make a serious effort to probe the cause.

Such arguments provided food for thought, but amongst television journalists themselves the discussion usually boiled down to a more straightforward question. Was it possible to cover current affairs seriously and still attract a significant audience? Commercial companies could not afford to slot high-principled but unpopular programmes into their peak-time schedules, as the BBC could, and that was why, many felt, the quality of documentaries on ITV was falling. Was it inevitable that serious programmes should mean small audiences? Not everyone thought so. In America, after all, a programme called '60 Minutes', serious and conscientious though relying heavily on brash exposé techniques, was nearly always among the top-rated programmes of the week (although its timing, early on Sunday evening, may have had something to do with that).

The three men in the room believed fervently that a mass-appeal news-based programme was achievable, although they doubted whether the commercial companies or the BBC would have courage enough to try it. That was why many of the ideas for reform had raised the need for a 'third force' in news, unbound by the conventions of the existing networks.

The trio went over the familiar arguments. It was Martin Smith who asked: 'What about breakfast?'

Early morning television was popular in America, Australia and parts of Europe, where it was assumed that viewers wanted a news show at that time of day. It had only been tried in Britain in 1977, when Yorkshire Television, co-operating with the IBA, had undertaken a nine-week experiment. It had not been a success but that was scarcely surprising, since the programme had been unoriginal and the timing wrong – 8.30 to 9.30, when many of the potential audience had already left for work. Yorkshire had filled the hour with a short news bulletin, cartoons and an instalment of the American soap opera 'Peyton Place'.

The main reason that nothing more ambitious had been tried by Yorkshire, or anyone else, was the cost. The technical unions operated under so-called 'golden time' agreements by which they were paid enormous sums for working in the small hours. Under one of them, technicians would get up to 4½ times their basic rate after midnight. Members of another union, if they were still working after 1 a.m., would be paid for an extra 24 hours, no matter how short the excess period. No regular commercial operation, however successful, could absorb such costs.

Yet since any brand-new service would create extra jobs in all areas of television, the unions would presumably agree to negotiate if a firm proposal for an early morning service were made. Why not, said Smith, suggest to the IBA that they introduce a breakfast franchise?

Dimbleby was enthusiastic about the idea, Elstein less certain. But the discussion was stimulating enough to encourage Dimbleby, after it had broken up, to sit down and write a rough prospectus for a morning television service. Discussing it when they next met, they decided it was worth taking the matter further. But they recognised that the three of them by themselves would have insufficient clout with the IBA for any proposal of theirs to be taken seriously. They needed someone with a national reputation to impress the IBA with their seriousness, but someone with a fresh turn of mind who would see the possibilities in their scheme. The right man would epitomise the responsible but popular approach they envisaged for their station. They decided to contact Harold Lever.

Soon to become Lord Lever, he had been in the Cabinet throughout the five-year life of the Labour Government that had been defeated earlier in 1979. He was an economic adviser to Harold Wilson and James Callaghan during their premierships. A genial, undogmatic man of sixty-five, a popular member of the centre-left establishment, he seemed an ideal choice.

Lever thought so too, for he quickly agreed and the group – now expanding quite fast – began meeting in his ornate flat in Eaton Square. New people were invited to join and strengthen the group journalistically. They included Peter Jenkins, the *Guardian*'s political columnist, David Watt of Chatham House, Catherine Freeman, Mavis Nicholson and Nick Ross, later to be successful on the BBC's breakfast show.

Together they drafted a letter for Lever to send to Lady Plowden, chairman of the IBA. But Lever knows that things are not generally achieved by correspondence. Without telling the other members of the group, he arranged to go and have a talk with Lady Plowden. She was not encouraging. She said she saw little chance of the Authority approving the new television service he proposed. The professional staff, she explained, would be against it, largely because with the impending launch of Channel Four they would have enough on their plates already. But she did not want to deter him from writing. No harm in trying. It was a confidential meeting so Lever did not tell his colleagues about it. Not wanting to appear to be dampening the hopes of so enthusiastic a group, he let them go ahead with the draft and despatch of the letter. For some weeks Lady Plowden did not reply. When she did, she was no more encouraging than she had been when she saw Lever.

At the same time, another group was pressing the IBA on the matter of an additional franchise. For about a year Norman Strauss, formerly head of the policy unit at 10 Downing Street, had been recruiting people to discuss a possible initiative in television. They included Hugh Stephenson, editor of *The Times* Business News; Adrian Ball, formerly with the commercial radio station London Broadcasting; Jerome Kuehl of the National Film School; and Roger Graef, a leading maker of current affairs films for TV (he later dropped out). Like the Lever group, they were all concerned about the quality of communications in general and television in particular, but they viewed it from a rather loftier perch. Noting that little had changed in television technology for some twenty years (bar colour), they could see major innovations round the corner. Satellites, video recorders and optical cables would turn the television set into a flexible video display unit. They argued that a company not locked into existing TV franchises would be more likely to exploit such changes beneficially than a company whose experience was in broadcasting across airwaves.

Their proposal was that London, by far the largest market in Britain, should be split three ways instead of the present two, where one company provides the programmes during the week and another at weekends. The third company, they advocated, should be allowed to transmit in the daytime, when the other two were less interested anyway. They wanted to start at dawn and hand over to the evening contractor at about 5.30. The programmes would be mostly documentary and educational. The group forecast advertising revenue of £20 million a year and would have used the programmes as an element in multi-media publishing.

They wrote to Lady Plowden and, for good measure, to *The Times*. The IBA chairman's reply was couched in similar negative terms to those she had used to the Lever group. Strauss and Stephenson had argued that it was time for new blood in commercial TV. She responded that there was an equally forceful case to be made for those companies that had risked investing in television to be allowed to enjoy the fruits of their investments.

The IBA is a unique, peculiarly British body. It was created in the early 1950's as the Independent Television Authority, to mollify those who feared that the introduction of commercialism into television would mean an inevitable collapse of standards. Its members are for the most part unconnected with television except as viewers, and are chosen for their probity and their standing in the community. They have one over-riding sanction against companies whose programmes fail to come up to scratch – they can simply take their franchise away at the end of the eight-year term or, in cases of extreme provocation, at whatever time they wish. Yet equally their powers are compromised by the realisation that the companies they regulate are first and foremost commercial concerns, their chief objective to make a profit. A high level of capital investment is required to start a TV station. People whose own money is at risk will in the end do things their way. Although the IBA can insist on a measure of serious and less popular programming, it cannot ultimately overcome this commercial imperative. Its relationship with the companies is therefore ambivalent and has never been properly resolved.

In the autumn of 1979, to prepare for offering the new franchises, members of the Authority and senior staff decamped for a weekend to Great Fosters, a large road-house near the Thames at Egham, whose most memorable decorative feature is a collection of pictures of Ginger Rogers, the musical comedy star, in the lobby. (She once stayed there.) But gazing at her starry twinkle failed to inspire, at least in the staff, any sense of sparkle or adventure. When the question of a new franchise was broached they delivered dire warnings. It would divert the attention of potential advertisers from Channel Four, which, being complementary rather than competitive with the main ITV channel, was anyway destined to attract only a minority audience and could find it hard to make ends meet. Moreover, twenty new commercial radio stations were due to be established, and the advertising for those had to be won as well. Finally, look what happened to the Yorkshire experiment.

While members of the Authority agreed quickly that the Strauss/Stephenson London daytime scheme had little to be said for it, several were keen to explore breakfast television further. Lady Plowden was conscious that the IBA would be accused of being dull and conservative if it just handed out the same old franchises. She had powerful support in Christopher Bland, a businessman whose philosophy is that if a group out there is prepared to take a commercial risk, why should the Authority prevent it? But Sir Brian Young, the Director-General, and Colin Shaw, head of the television section, spearheaded the opposition, and they won the day. It was decided that breakfast TV was not to be. That was when Lady Plowden sent what amounted to a brush-off to the Lever group.

Christopher Bland, whose final year on the Authority this was, did not regard the Great Fosters decision as final. A few weeks after that riverside weekend, the IBA had to travel to Manchester for one of their regular regional meetings. It was on the train back to London, in the convivial atmosphere of a first-class compartment quite late in the evening, that Bland raised the breakfast question anew. He declared to anyone who would listen that the IBA's plans were frankly boring. What was it they would have to announce the following January, when they were formally to invite companies to apply for the regional franchises? There was to be a small adjustment in the Midlands but very little else. For the rest, it would be the same old franchises and, in the end, most would probably be awarded to the same old companies. Hardly, he told Lady Plowden, a very adventurous note on which to end her term as Authority chairman. And he repeated the argument he had made at Egham: if entrepreneurs were prepared to take the risk with their money, who was to say they should not? It might be that if a breakfast franchise were offered nobody would want to operate it. So be it. That would be a commercial decision and the Authority could not do anything about it. But at least businessmen should be allowed to weigh up the possibilities and see if they wanted to take the risk.

These were powerful arguments, cleverly designed to exploit the consciousness of IBA members of the ambiguity of their relationship with commerce. Sir Brian Young, in particular, was acutely aware that he and his colleagues were bound to be criticised whatever they did. He had been through it all before. When the time came for reallocation of the franchises, the choice was between leaving things much as they had been and making wholesale changes. If they took the first course they would be accused of croneyism, of being too friendly with the existing companies and therefore unable to judge their performance with critical objectivity. If they got rid of more than one or two they were charged with making changes and introducing instability for the sake of it, of acting on an irresponsible whim. Given that they probably were not in fact going to make many changes in the regions, Sir Brian could see that to launch something as eye-catching as breakfast TV could well divert that first strand of criticism. By the time the train had reached Euston, he and some of the other officials began to moderate their previously firm resistance to the idea. Perhaps there was something in it after all.

The view gained ground in the ensuing days. At their December meeting, the Authority reversed their earlier decision and resolved that they would invite tenders for a breakfast franchise – but they would do it guardedly, leaving themselves

45

the option of not awarding the franchise if no suitable application was made, or if for any other reason it was deemed impolitic. Officials made hurried phone calls to Lord Lever. Could he get his group together and clarify a couple of points on their proposal? Rather baffled by this sudden about-face, he nonetheless did what was asked.

On 24 January 1980, Lady Plowden announced details of the contracts that would be awarded that year to begin in 1982, and for which applications were invited by 9 May. The breakfast proposal was appended almost as an afterthought, which indeed it had been, and it was couched in tentative terms. 'If a breakfast-time contract was awarded, it would apply between 6 and 9.15 a.m., seven days a week', although in practice it was thought the programmes would not begin until 7 a.m. The service would contain chiefly news, information and current affairs, in a pattern that had been established with the 'Today' programme on Radio 4. The existing news organisation, ITN, would be expected to play a role as the supplier of some of the news material and applicants would be asked about their proposed relationship with them. Did the Authority sense trouble even then?

The IBA's announcement caught the imagination of the press. There were cartoons about the exciting new prospect (new in Britain, at least) of keeping your eyes glued to the screen while chomping away at the cornflakes. On a more serious level, an editorial in *The Times* gave the idea 'a modest welcome', though it added: 'The need to extend the hours of television coverage is not immediately evident.' And it gave a warning of danger if the BBC was allowed to enter the morning slot too. Any increase in the number of channels competing for viewers 'is liable, unless there are stringent safeguards, to reduce the general level of quality in the frantic search for audiences'. That argument was prescient, but it has a corollary. If you deny the possibility of competition you establish a monopoly that is in effect giving the public what you think they should have rather than what they want. That dilemma was at the heart of the argument about the creation of commercial television in the first place, although then it was the BBC's monopoly that was at issue, instead of a commercial monopoly that would be threatened if the BBC decided to compete.

The Times, was the first newspaper to grasp what the offer of a new franchise would mean in terms of the people working in the industry. The period leading up to the re-awarding of the regional contracts is always rife with rumour about which key staff from the BBC or an ITV company are being lured to form part of a group making a bid. Now, with something completely new on offer, there would be many more such attempts at competitive enticement. On the day after the announcement was made David Hewson wrote in *The Times* that the BBC and the independent companies were looking out for 'moles' secretly doing deals with franchise hopefuls. He quoted the established companies as saying that anyone caught associating with one of the new groups would face disciplinary action or even the sack. But that was not an effective deterrent. Between then and 9 May, when the bids had to be in to the IBA, few worth their salt in current affairs television had not been approached by one of the hopefuls; and many of them succumbed to the lure. Hewson, reporting that the IBA's move had been made after prompting by the Lever/Dimbleby group, wrote: 'Waiting in the wings are a number of other journalists and broadcasters, some of whom are well known to the public and others little known outside their professional circles.'

Michael Rosenberg is a lean, dark entrepreneur, generally sporting a handsome tan, who conducts himself with an air of confidence and exuberance. His main business is United Medical Enterprises, a private medical company with interests overseas. But since the early 1970s he has also been a close friend and business associate of David Frost. He has shareholdings in David's production companies (called the Paradine companies because Paradine is David's middle name). David relies on him for financial advice and they talk regularly on the telephone.

In January 1980 Michael Rosenberg was reading his paper on a flight from London to Los Angeles. (Like many modern business sagas, a good part of this story takes place in the air.) He saw the report of the IBA's breakfast initiative and his thoughts turned immediately to David, also in America. David had been a television star on both sides of the Atlantic for the best part of twenty years. He first came to public attention in the BBC satire show 'That Was The Week That Was', soon after he left Cambridge. He was the show's presenter, the public face of a team of lively wits – some, like him, veterans of the Cambridge Footlights Revue. The show set a new fashion for sharp, penetrating comedy and David capitalised brilliantly on the fame it brought him. He starred in a succession of other funny series, then went to New York to help establish similar shows – and himself as a small-screen personality. For a period, when appearing in both countries at once, he would fly across the Atlantic twice a week.

He had always been interested in the entrepreneurial side of television as well as in appearing on screen. In 1966 he played a leading role in putting together a consortium for broadcasting to London at the weekends. It had won the franchise largely because of the big names David had attracted, notably Michael Peacock, controller of BBC television, former junior minister Aidan Crawley, script writer Frank Muir, then head of BBC light entertainment, and Peter Hall, co-director of the Royal Shakespeare Company. That was a golden period for David. Still only in his twenties, he had become one of the most familiar and popular figures on the British screen. He was making a name for himself, too, as a confidant of the powerful. In 1966 he hosted a breakfast party at the Connaught Hotel attended by Harold Wilson, the Prime Minister, and numerous celebrities. For a man who had already become part of the media Establishment at that tender age, who knew what the future might bring?

When London Weekend suffered a financial crisis not long after it went on air, the unsinkable David emerged unscathed, putting it down to experience and to the teething troubles inevitable in getting any new organisation off the ground. After playing a role in securing the appointment as chairman of John Freeman, former British ambassador in Washington, David soon gave up his interest in the company. But his Anglo-American television career continued to flourish. A notable coup was obtaining the first long interview with ex-President Nixon after his disgrace in the Watergate affair. And his production company was involved in the popular BBC comedy show 'The Two Ronnies'.

That was why Michael Rosenberg thought immediately of David as he read about the coming of breakfast TV. When he arrived in Los Angeles, almost the first thing he did was to get him on the phone. But he did not receive the enthusiastic response he had been hoping for. David was absorbed in one of his other projects and said simply: 'Yes, yes, we must talk about it,' in a tone that suggested he had scarcely taken the idea on board. Michael waited a few weeks before broaching the subject with David again. By this time David had himself seen the press articles about breakfast TV and was beginning to take more notice. But still nothing firm was agreed.

The next stage had to await another intercontinental air journey. Lord Marsh, chairman of the Newspaper Publishers' Association, was deputy chairman of Rosenberg's United Medical Enterprises. In March they had to visit the Bahamas together on company business. As they cracked the second bottle of champagne in the first class cabin, Michael told Dick Marsh of his interest in the possibilities of breakfast TV and of his so far inconclusive approach to David on the matter. Dick was immediately interested. Since leaving active politics and then ending his stint at British Rail, he had marketed himself as a figure of public standing who would grace any boardroom. He had also gained, quite late in life, a degree of business acumen. A place on the board of a television company would, he felt sure, be appropriate and enjoyable for a man in his position. The flight to the Bahamas takes eight hours and by the end of it the two men were convinced that it was a tremendous idea. They discussed it further as they lay side by side on the sunny beach, their business having been concluded swiftly. They even took soundings – fruitlessly, it turned out – to see whether local businessmen would help finance the venture. When they returned, Michael and David talked for a third time.

The involvement of Dick Marsh made the project seem that much more tangible to David, who said he would go ahead if one further condition was fulfilled. They must, he said, approach Peter Jay and get him involved at a high level. David knew how the IBA worked and was convinced that Peter was exactly the kind of figure the Authority would award the contract to. David had been to visit Peter a few times at the Washington embassy and they had naturally talked about television. One of the discussions, David remembered, had been about how good the American morning programmes were and how it was a shame nobody had tried anything similar in Britain. So far as David knew, Peter had no pressing commitments just then and might with luck be able to devote considerable time to the breakfast venture. David undertook to contact Peter right away.

Peter was extremely interested in breakfast television and had already been approached by others who shared David's view of his suitability to head a bid. One of them had been from the Norman Strauss/Hugh Stephenson team who, since their idea for a London daytime franchise had been specifically turned down by the IBA, had decided to bid for breakfast instead. The other was from a group that included Charles Wintour and Jocelyn Stevens of Express Newspapers, with Lord Grade's Associated Communications Corporation, who planned to base the station at Grade's film studio at Elstree. When he heard David's proposal Peter was quickly interested, both because he knew and liked Frost and because he saw a large role for himself in planning the enterprise and creating the infant concept.

Although it took six weeks for Peter to agree finally to join, David had the strong impression from the start that he would do so and went ahead with his planning. As it happened, he found himself in America not long afterwards with Michael Deakin, making a programme about Elvis Presley for Yorkshire TV. When David mentioned the idea to Michael and asked whether he would be interested, the response was immediate and positive; and soon Michael hit on the United Artists concept. When David put this to Dick Marsh and Michael Rosenberg they were enthusiastic, especially Dick. He believed that it would clearly distinguish their application from the others. It was already clear that there would be a good many strong bidders for this franchise and he guessed that those who offered innovations in the structure of the company as well as in the programming might gain the edge. If any of them foresaw that an artists' co-operative might be hard to work in practice – well, let's get the contract first, then worry about that. This was, after all, a novel franchise which had to be approached in a new way.

Time was running short. By the time Peter said yes, at lunch with David in New York, there was little more than a month in which to recruit others, to find a studio and, most important of all, to look for the £10 million of financial backing the project would need. Then they must draw up the detailed application document required by the IBA. Their slow start had already cost them their first choice of name for the new company. Their original thought had been AMTV but that had already been spoken for by the Lever group. So they settled for TV-AM, later altered typographically to TV-am.

David was the obvious candidate for recruiting officer and his method was simple. He approached, one by one, the best-known names in television news and current affairs, offering them large salaries, a slice of the equity, and only six months' work every year. The first to be signed up was Michael Parkinson, the journalist turned TV personality who had become host of a successful BBC talk show. His agreement to join was vital in convincing David the project could work. Next was Angela Rippon, whose canny sense of her own worth as a personality, backed up by a smart agent, had led to her broadening her activities from newsreading to public appearances, commercial endorsements and even to writing a book about the equestrian career of Captain Mark Phillips, husband of Princess Anne. She was won over at an expensive dinner at Le Cygne in New York. Robert Kee was brought in as the most respected of the older generation of current affairs specialists, a veteran of 'Panorama' and the man who launched ITN's mid-day news programme.

Esther Rantzen offered some resistance. When first telephoned by David she was not at all enthusiastic. She assumed, to begin with, that the franchise would be awarded to ITN, who were putting in a strong bid. And she was happier with her weekly programme than she had been doing daily reporting on the BBC's 'Nationwide' some years earlier. With a young family (her husband is former BBC executive Desmond Wilcox), she was not at all sure she wanted to be getting up to go to work in the middle of the night. But David was a friend, so she asked him to her house near Kew Gardens, to talk about it.

Strolling in her garden on a pleasant spring afternoon, David told her for the first time that Peter was going to be chairman. He also revealed that Nick Elliott, a respected executive of London Weekend Television, would be Director of Programmes. Esther was impressed by those two names and enticed by the prospect of a share of the equity. Nevertheless it was not until David brought Peter Jay to Kew in person, and sat him at her dinner table to switch on his full powers of persuasion, that she succumbed. It was ironic, in a way. On numerous occasions in 'That's Life' Esther has warned viewers that they must not trust smooth-talking salesmen who insinuate themselves into the house and start making their sales pitch. True, Peter and David were salesmen of a very superior order, but the principle was the same. Yet in the event, Esther had, after all, learned from her own programmes. For what she always stresses to viewers is that nothing should be signed in haste that cannot be unsigned at leisure: there must always be a get-out clause. A few months later she was to be glad of the get-out clause in her own arrangement.

Peter and David have always insisted that the original presenters were selected not just because they were among the most famous faces on television. As it happened, they were; but more important, it was claimed, was that they all had considerable experience in producing as well as presenting programmes. That was demonstrably true in the cases of Esther, Michael Parkinson, Robert Kee and David himself, although Angela's feats of production were less well known – as were Anna Ford's, about whom a similar claim was made when she joined later in the year. They were all expected, it was said, to 'have input' into the programmes, a comfortable piece of jargon that had the virtue of not meaning anything specific. That promise was what lured some of them to join. It was stressed both in the TV-am application document and the interview at the IBA at the end of the year. It was a benign, enlightened idea. Nobody guessed it carried in it the seeds of disaster.

At that early stage there was only one incident that marred the ebullient optimism of the spring days. Michael Deakin had assumed that, as one of the very earliest founders of the consortium, he would be director of programmes, the top job on the programme-making side. Yet Nick Elliott, who had worked with Peter on 'Weekend World', felt that *he* should have the title. It fell to Peter to arbitrate and devise the *primus inter pares* solution which in effect established two separate chains of command. Nick would be director of programmes. Michael's role would be director of features, and they would each run separate empires. Michael was uncomfortable at having to take what, however it was packaged, amounted to second place. But he reminded himself that the chance of their winning the franchise was remote, so the question would become academic.

Finding a studio was another requirement that had been rendered harder by the late start. Rival bidders, moving faster, had snapped up most suitable existing studios near the centre of London. There was nothing to stop any of them proposing to build a new facility of their own, but if they chose that course they would have to prepare for the IBA, as part of their application, a list of all the equipment they would provide. The IBA sets high technical standards for its member companies. Given the time available it was preferable to find a studio and simply list the bulk of the equipment in place, rather than sketch out from scratch an entire new studio plan.

Someone suggested Ewart's, a modern facility in Wandsworth, in southwest London. David Frost and Tom Cook, the group's technical adviser, went to see its owner and founder Keith Ewart. A few days later Peter went to view the place and so did Dick Marsh. Keith was reluctant to commit himself firmly in writing and refused to sign a letter David had drafted for him. But he agreed that his studio should be identified to the IBA as the one from which TV-am broadcasts would come. On 18 April Dick wrote to Ewart welcoming the agreement to work together. 'Quite apart from the purely business side, I think it should be both fascinating and fun,' he observed. Later he was to suspect that Ewart took the 'fun' part too literally.

Michael Rosenberg was put in charge of raising money. Very rough calculations had produced the conclusion that £10 million was probably needed as start-up capital. Pondering where he might start to look, he received a surprise phone call from Michael Cumming at Barclay's Bank. The two of them had worked together in the days when Rosenberg was in banking. Cumming said he had been approached for financing by an applicant for one of the regional TV franchises. Knowing Rosenberg's connection with David Frost, he was calling to inquire what he knew about the potential client. Rosenberg helped him as best he could but the call set his mind working. Barclay's were obviously interested in financing television projects. So he rang another friend, Oliver Stocken, the director then responsible for Barclay's Merchant Bank. After a series of meetings, Barclay's agreed both to take up a shareholding and provide a loan.

Just over a third of the equity was reserved for the founders – Jay, Marsh, Rosenberg, Deakin, Elliott and the five presenters. They had to put up their own money to buy it, but the terms were highly favourable to them. The founders between them paid £250,000 for £200,000 worth of stock – only £1.25 per share – while the investors produced £875,000 for a nominal £350,000 worth, or £2.50 per share. The investors also put up £4,375,000 for subordinated loan stock. Barclay's gave a loan facility of £2.5 million to bring the capitalisation to £8 million, while the investors undertook to provide a further £2 million if necessary.

The final housekeeping detail was the composition of the board. For the most part it chose itself, from representatives of the investing companies. It was agreed, though, that what it lacked was a woman. Lady Plowden, the chairman of the IBA, was thought to set great store by the participation of women in such enterprises. As the team racked their brains for someone suitable, it was Dick Marsh who came up with Jennie, Lady Enfield, a vivacious country lady with many public duties centred on her home near Winchester. Dick had met her when he spoke at a political dinner she had organised. A Tory, though not a high Tory, she would correct any impression of centre-left bias on the board. She was excited to be asked and quickly agreed to join and to buy a token shareholding. She was about to divorce Viscount Enfield and in 1981 married Christopher Bland, the breakfast TV protagonist who had fortunately now left the IBA board.

All that remained was to compose the all-important application document that had to be in the hands of the IBA by 9 May.

Most of those involved gathered at David's Knightsbridge house on the May Day holiday weekend to draw it up, on the basis of guidelines set by the Authority. It was a hectic and exciting weekend. Rosenberg and the bankers were in the ground floor dining room, sitting round the polished oval table to draw up the financial part of the document. Peter, David, Deakin, Elliott and the other programme people were in the first floor drawing room drafting the section on programme intentions. Tom Cook and his technical people had commandeered the kitchen. Secretaries were typing in the downstairs office and in the small sun lounge on the half landing between the ground and first floors. If anyone came to see David – he has a stream of visitors on most days, to discuss this project or that – the only place they could go was the bedroom.

The gang were pleased with the finished product. But one question in the IBA's list that neither they nor any of the other applicants could properly answer was what relationship they proposed to have with ITN. The snag here was that ITN were also bidding for the breakfast franchise. They thought they were ideally placed to provide the reliable and lively news, features and commentary envisaged by the IBA. They pointed out, too, that running a twenty-four hour operation would put them on a par with the BBC and the American networks. They were not impressed with the argument for a 'third force' in news; nor did they worry that, because ITN is jointly owned by the regional ITV companies, giving them the franchise would effectively give it to the same people who controlled the network at other times of day: cronyism.

Because they were serious bidders themselves, ITN refused to start discussions with the other contenders about the provision of a news service. Some of the groups were not planning to have anything to do with ITN anyway, but those who did envisage a relationship could say only what TV-am said: 'In the event that ITN services are available, TV-am would wish to enter into detailed conversations with them, since they would like to use a number of short ITN bulletins. . . . Since TV-am have among their key personnel presenters with a background in newscasting, they would also like to discuss with ITN the use of their own personnel as part of the news presentation, as they believe that a melding of news, current affairs, feature and general material into one free-flowing programme is of the utmost importance to a successful broadcast.'

Had they been asked which paragraph in the whole document would come back and haunt them later, the gang would not have chosen that one: but it did. For the time being, they were delighted with their work. Peter showed it to his old mentor John Birt, who thought it a winner too. The only small cautionary voice was that of Jonathan Aitken, who told a friend as he read the document: 'Gosh, I hope this isn't going to be like G. R. Elton's description of Cardinal Wolsey's foreign policy – splendid in design but futile in accomplishment.'

On 9 May, when the application was safely delivered to the IBA, David gave a party at his house for those who had taken part in writing it. The atmosphere was easy, cordial and high-spirited. They were all such good friends. The air was electric with mutual inspiration and team spirit. It was no surprise that Esther and David and Michael Parkinson, all old troupers of the small screen, should hit it off: but even the money men seemed to fit in perfectly, despite Aitken's scepticism. That frenetic weekend when they composed the document had moulded them into a cohesive unit. It would be a pity if it were to break up. Emotionally, some of them agreed that even if they did not win this franchise the group should stick together and try some other endeavour – like people in their last term at school arranging to meet in five years' time. But David and Peter, those two indomitable optimists, were growing more and more convinced that they were going to succeed.

Eight groups applied for the breakfast franchise. This was many more than the IBA had expected and made it effectively impossible for them to say that they would not now grant it. During February Christopher Bland, the authority member most keen to see breakfast TV introduced, did some discreet touting to get people interested. His term on the IBA had ended in December but he was still anxious for the initiative to flourish. Had it not been for the rule that forbids any former IBA member becoming active in commercial television for two years after he quits, Bland might have been tempted to get involved himself. In any case, he did not want the Lever group – the only one then known to be applying – to win without a contest.

Bland spoke to Christopher Chataway, former Olympic runner, newscaster, Member of Parliament, junior minister and now a successful banker. Next he approached James Lee, the young deputy chairman of Pearson Longman, the publishing conglomerate, and brought the two tycoons together. They decided to make a bid. Through Lee, they recruited the redoubtable Harold Evans, probably Britain's most accomplished journalist, then editor of the *Sunday Times*. Other directors named included Liz Calder, editorial director of Jonathan Cape, and Mike Wooller, head of documentaries at Thames Television. Evans was not identified in the application document but he would have been managing director of the new company, to be called AM Television. He also wrote most of the application, ensuring that it was the best written of all of them. It began with a quotation from Shakespeare's *Henry IV, Part 1*:

> Owen Glendower: I can call spirits from the vasty deep.
> Hotspur: Why, so can I, or so can any man. But will they come when you do call for them?

The application went on to explain why, in Evans's opinion, the viewers would come when AM Television called for them. There were some charming touches. A short item was planned every morning containing the 'good news' customarily buried underneath reports of catastrophe in regular bulletins. And there was one paragraph that, to those who know him, could only have come from Evans, author of *Newsman's English*, a standard work for journalists:

> We lay particular importance on the quality of the writing and editing. In set pieces on AM Television, viewers will not be told about 'the unemployment situation' but that there are 1,390,000 people out of work; and when it is snowing harder outside that is what viewers will be told, not, as so often in broadcasting verbiage, that 'weather conditions are deteriorating'. The morning

mind above all does not relish being stuffed with cotton wool prose.'

Like many of TV-am's rivals, AM Television took a gentle swipe at the 'famous five' concept in their application. 'We expect that no more than two, perhaps only one of our five principal on-screen presenters will be familiar faces,' Evans wrote. 'We have aspirations so far not reflected on British television and we acknowledge to ourselves the vital need of discovering the new style of presenter we have described. We regard the discovery and nurturing of new talent as rather neglected and yet essential for the development of television in Britain.' The group would not have used ITN as a news source at all. They would have operated from a Soho studio run by Trilion Video.

Another serious contender was Morning Television Ltd., whose big names were for the most part solid production people with one exception, also a closely-kept secret: David Dimbleby, Jonathan's brother, would have been the chief presenter. Robin Scott, former managing director of BBC Television, was the chairman, and the managing director would have been Mike Townson, editor of 'TV Eye' for Thames, later to play a role in the TV-am saga. The group planned to sell shares on the open market as soon as they won the franchise, to preclude any loss of control to the financial institutions. They would have used ITN for foreign news and had also located a Soho studio, run by Molinaire.

AM Television, Morning Television, TV-am and AMTV (the Lever group) were seen in the business as the four likeliest contenders. The IBA had virtually decided that ITN should not have the franchise. The argument about plurality of news sources weighed heavier with them than ITN's point about needing twenty-four-hour exposure. The ITN application was, it was true, backed by the existing programme companies. (John Freeman of London Weekend was their eloquent spokesman at the December hearing.) The regional companies would prefer the news service to be run by an organisation in which they had a share rather than by a new company that would be competing with them for advertising.

The three remaining applicants could only be regarded as outsiders. Daytime Television was the group that had advocated an all-day franchise in London. Even strengthened by the formidable Establishment figure of Baroness Trumpington as chairman, this group was seen by the IBA as being top-heavy on the intellectual side, probably unable to spice their programmes with the levity needed to win and keep viewers. Their application contained its own version of the Jay mission. They would 'explain the relationship and relevance of news that has happened to what is about to happen and to the lives and aspirations of viewers themselves.' Yet they criticised the Jay-Birt theory specifically. 'Insofar as this school of criticism implies that there is some single objective truth in news and current affairs which TV could put across to its audience, we find that criticism both unconvincing and elitist.' And there was an additional implied assault on TV-am's big-name presenters: 'The nature of TV at breakfast time will require individuals with whom the audience can identify, rather than individuals whose contribution is their own obvious star quality.'

Daybreak Television was the Charles Wintour and Jocelyn Stevens group, linked with Lord Grade's Associated Communications Corporation. Some of the names attached to this bid were quite impressive. Francis Essex, director of production at ATV, would have been managing director. Alan Whicker was involved, and so was the former England cricket captain Mike Brearley.

Finally came Good Morning Ltd., the least news-oriented group of all. Only Julian Pettifer, a respected reporter, carried any weight on the news side, but they were strong on entertainment, having been created by Terry Connolly of Chrysalis, the record and pop music company. Tim Rice, the lyricist, was with them, as well as medical journalist Dr Miriam Stoppard, and the witty Ned Sherrin. The only novel feature of their application was that they wanted to do pure entertainment – game shows and the like – between 9 and 9.30 a.m., to increase their revenue.

When all the applications had been read, the big names noted, and the other even bigger names – the anonymous ones – gossiped about, it seemed that nearly everyone who was anyone in current affairs television had been roped in – with the possible exception of Sir Robin Day, who was approached by at least two groups but joined none. As Ned Sherrin put it: 'Bidding for a TV station is a bit like having a boutique in the sixties – you don't like to be left out. You have to be on board somewhere or other just for the sake of the memoirs.'

Members of the IBA took the eight applications, as well as dozens of others for the regional franchises, to read on their summer holidays. They would hold hearings on them when they returned, with those about breakfast television scheduled for December, not long before the decisions were due to be announced. In the intervening period most companies campaigned to advance their claims. They held press conferences to promote the idea first of breakfast TV in general and then their particular qualifications for running it. The first object was achieved by commissioning a series of studies from research organisations about morning habits. Statistics abounded but nobody was quite sure what to make of them.

'The astounding thing,' said Robert Worcester of MORI Polls, 'is that 88 per cent of people don't eat breakfast in the dining room.' It depends what astounds. ITN's researchers discovered that 24 per cent of people had their main TV set in the room they had breakfast in and that 46 per cent of people questioned said they would watch breakfast TV.

Meanwhile the IBA itself was producing research results that suggested breakfast television might be a misnomer. They concentrated on where second sets were placed. Although only 30 per cent of their sample had second sets, more than two-thirds of those were sited in a bedroom and less than a third in the kitchen. So perhaps it should be called bedside television. If so, would the type of programmes thought suitable need to be reconsidered? The IBA researchers also produced another disturbing report, based on how viewers reacted if you increased the availability of certain categories of programme. In most cases, if you produced more of a particular type of entertainment – drama and adventure series, for instance – the viewers would watch more of them. But if you produced more news they did not usually view any more of it. This suggested that the

appetite for news was finite: once satisfied, people sought
entertainment and interest elsewhere.

As for whether anyone actually wanted morning television,
the most telling observation came from Ned Sherrin (who
seemed to enjoy the role of sceptical chorus to these events,
maybe because his group had scant chance of winning the
franchise). 'A substantial majority,' he observed, 'never asked
for the avocado pear. But once it started to come in they were
frightfully keen on it.'

The IBA was meanwhile going through its own peculiar form
of democratic consultation. Of the two public meetings, that
at Croydon was most notable for Peter Jay's invention of the
'mission to explain'. It was one of Peter's more successful
speeches. He spoke last of the eight group leaders and was able
to pick up points the others had made. When he sat down
Jonathan Aitken, in the audience, applauded ferociously. The
young woman in the next seat tapped him on the shoulder. 'Are
you his brother, by any chance?' she enquired.

The meeting at Darlington is best remembered by the
participants for the dreadful weather and for a singular
contribution from a member of the audience. It took place in
a room at the Kings Head Hotel near Darlington railway station.
It was rainy and windy. Executives from the Charles
Wintour/Francis Essex group decided to save time and go by
helicopter, but the weather delayed them and they arrived
midway through the session. One of the few non-television
people present was a man who announced himself as
representing a local cycle road racing club. The club's
committee, he declared, had discussed breakfast television and
decided they were against it. This was not just because they
felt disinclined to switch their sets on at that time of the morning
but because of the children. They observed that children will
watch almost anything on television and would probably be
particularly keen on the commercials. Lingering in front of the
screen, they would delay their departure for school, be forced
to hurry, and be maimed in road accidents.

Ludicrous as that suggestion may have been thought, it at
least provided relief from the worthy speeches coming from the
representatives of the contending companies. After the hearing

the galaxy of media people descended on the dining room. A few, whose involvement in a bid had not been formally announced, were forced to assert unconvincingly that they were just there to listen.

The IBA set aside two days in December to hear the claims of the eight contending groups. They had read the application documents, written to all the groups with supplementary questions and read the answers to those. They had obviously reached in their own minds some tentative conclusions, but few doubted that the impression made at the oral hearing, the *viva voce* test, would be decisive.

As professional performers, for the most part, it was inconceivable that the applicants should attempt the ordeal without rehearsal. A few days before their appearance, therefore, Peter and his TV-am colleagues gathered in a private room at the Carlton Tower Hotel, Sloane Street, for a mock examination. The judges were John Birt and Professor Tom Carbery, who used to represent the Scottish interest on the IBA, so had experience of the kind of questions asked.

The rehearsal was held, by unhappy chance, on the day John Lennon, the former Beatle, was shot dead in New York. That upsetting event may have helped account for the abysmal performance Peter and the others put up. Birt was scathing in his criticism, especially of Peter, who was taking most of the questions himself and being unbearably long-winded. They were all being too meticulous about answering the questions, instead of using them only as a lead-in to points they wanted to make. The object was to put across their message and concentrate on making that pitch, no matter what questions were asked.

Carbery was no more gentle. 'Look at you,' he mocked, gesturing to the forlorn group. They were – sin of sins – sitting back in their seats, like schoolchildren trying to make themselves invisible in case the teacher asked them a question. They ought to be perched on the seat edge, alert, interested and eager to participate. The contenders, taking their cue from these blunt criticisms, now began to snap at each other. It was a depressing moment, the first thing they had truly done together – and they had done it poorly. Belatedly it was dawning on them that euphoric self-confidence and affection for each other might not after all be enough to see them through. They needed to do some hard work and some hard thinking. As they left the Carlton Tower they felt deflated. Was the unthinkable possible? Might they not win the franchise after all?

IBA hearings are conducted with elaborate ritual, partly for practical reasons. It is not easy to get groups of a dozen or so people in and out of a central London building, at hourly intervals, and maintain security and order. Security is essential because at these meetings those members of consortia who opted for anonymity because of their other interests, are forced to present themselves before the board. Harold Evans, for instance, was actively engaged in forming a consortium to try to buy his *Sunday Times*.

The groups are ushered in through the door from the underground garage and taken up to the eighth floor, to the first of a series of rooms, each one nearer than the last to the board room where the hearings are held. One applicant likened it to visiting a powerful monarch in an earlier age, where in moving from one anteroom to the next the supplicant felt he was gradually getting closer to the divine presence. When the interview was finished they left by a door on the other side of the room and were whisked back to the basement. In that way no group saw any of the others.

A way of conveying the disparate, arbitrary nature of the IBA is to list its membership. These are the people on the Authority in 1980 who awarded the breakfast franchise:

Lady Plowden (chairman), a stalwart of numerous committees of enquiry, government bodies and other public activities;
Lord Thomson (deputy chairman), former Labour Cabinet Minister;
Lady Anglesey, a member of the Arts Council and the National Federation of Women's Institutes;
Mr A. M. G. Christopher, General Secretary of the Inland Revenue Staff Federation;
Mrs A. M. Coulson, a member of the West Midlands Health Authority;
Mrs J. McIvor, a lawyer from Northern Ireland;
Rev. Dr W. J. Morris, minister of Glasgow Cathedral;

Prof. H. Morris-Jones, Professor of Social Theory and Institutions
at the University College of North Wales;
Mr A. J. E. Pursell, managing director of Arthur Guinness;
Prof. J. Ring, Professor of Physics, Imperial College, London;
Mr G. Russell, executive of Alcan Aluminium;
Mrs Mary Warnock, senior research fellow at St. Hugh's College,
Oxford. (She had tutored Peter Jay in philosophy for a term at
Oxford.)

The Glaswegian Dr Morris is known as the wit of the group.
Once Lady Plowden was complaining about how long people
took to say things at hearings and wishing she had a red light
to switch on when they were running too long. Said Dr Morris,
'I'll see if we can arrange some claps of thunder for you,' and
everyone laughed. This time, as the glittering TV-am crew filed
into the room and arranged themselves along one side of the
oval mahogany table, opposite the Authority members, the Scot
remarked in a loud stage whisper: 'Och, I meant to bring my
autograph album with me but I forgot.'

The fame of the group was the subject of considerable
discussion at the hearing and some who were there feel that
it was by no means an unmixed advantage. Dr Morris's quip
may or may not have been seriously meant as a caustic comment
on what some saw as an attempt to dazzle the Authority with
established reputations. Lord Thomson took up the point in
a direct fashion. 'Can you convince us that you're more than
a star-studded group of personalities and might actually be a
television team?'

Peter was the first to leap to their collective defence. He
stressed that the programme philosophy had come first, then
the people chosen to fulfil it. They were programme makers,
not just presenters, who would help devise and prepare the
shows as well as perform. Their shareholding was a guarantee
of their continuing interest. They would be on the programme
committee, a vehicle for self-criticism and new ideas, although
final decisions would rest with the board and the executives.

Anna then explained that she had been approached by a
number of other consortia. 'I joined this one because I thought
they were the nicest, best qualified group of people to do the

job. I have arrived at a time in my life when I am lucky enough to be able to choose with whom I work. To choose to work with as exciting a bunch as this, with something totally new in TV, attracts me enormously. Also the possibility, which I've been told often before but which I knew to be true this time: "You'll be involved in programme making. Your ideas will get on screen." That's very important to me because I think one's brain as well as one's other talents needs to be used.'

Now Michael Parkinson joined the defence, but decided attack was its best form. He objected strongly to Lord Thomson's use of the word 'star', which he treated as some kind of slur. 'The assumption people have is that you're just a face and someone behind you speaks your voice for you. . . . We're all journalists. I'm a producer – I think we all have been in television, and directors.' He went on to agree with Anna. 'I like playing with first-division players and I think this is an incredibly strong team. . . . All of us here were born and raised in a television studio and are used to handling a live studio situation.'

Now it was Angela's turn. 'We're being offered the opportunity to set the standard for breakfast television for the next twenty years. This is a new breakthrough. If we were taking over an existing franchise we'd be taking over someone else's groundwork. . . . It's not just that we have our money in it but we're putting our reputations on the line. If you give us the franchise and we do not succeed our reputations are at stake.'

Esther was not at the hearing because she was making 'That's Life' for the BBC, but Robert Kee was later able to have his say. 'I don't think we can have – yes, I will say it – the sort of reputations we have in TV without having been rather good at communicating responsibly, without boring people.' The programmes would be about 'being culturally aware and having a dialogue with your viewers'.

The hearing started at 12.20 and Lady Plowden said it would last for about an hour. She had begun by asking Peter why he was especially interested in breakfast television. Peter, profiting from the session with Birt and Carbery, said it was the idea of doing television in the early morning that had brought the group together and 'inspired all of us to the excitement and enthusiasm we feel.' It was the next great frontier of TV and gave the opportunity to develop the new kind of television journalism that they believed was urgently needed.

'I don't want to take up much time talking about that. I think my views and some of my colleagues' are well known on that. We have – and I used this phrase at Croydon – this very strong dedication to what we call a mission to explain. We believe the world about us both at home and abroad is fascinating. We believe the traditional news media do not succeed in making it come alive, in explaining its relationships and making it meaningful to people. We believe that this is the supreme mission of good television journalism.

'All of us feel very strongly that this would give us the opportunity, not in a heavy way but in a popular and attractive way – in a way that would reach all the people but nonetheless in a new way. Nothing other than breakfast television at present provides that opportunity. It is, if you like, a journalistic mission within the confines of the definition which the Authority has given of a programme primarily to produce current affairs and information.'

That was clearly enough of that, and Lord Thomson chipped in quickly to change the subject. He asked the first of several questions on something that obviously and rightly worried the Authority – who was actually going to run the company? Peter had said that he would be chairman and that someone else would be appointed chief executive. Thomson asked whether he could be given a clue who it might be.

Peter dodged the question. He said that had deliberately been left open and it would anyway be easier to recruit someone worthwhile with the franchise safely won. The truth was that they had originally approached Bert Hardy, a former executive of Rupert Murdoch's News International who had played a pivotal role in the rescue of London Weekend. Although interested, he eventually joined the Lever consortium and quit that when made managing director of the *Standard*, London's evening newspaper. When he dropped out some thought it important to replace him. Jennie Enfield, in her letter to Dick Marsh formally accepting a seat on the board, had expressed her doubts about the lack of hard managerial experience among board members. Their limited business experience had not included the creation of new companies – a tough skill to acquire. Dick and Peter, both confident in their abilities, disagreed.

Now Peter, responding to another probe from Thomson, said he would be able to give most of his time to TV-am because he had no full-time job. His column on *The Times*, resumed when he stopped being an ambassador, might soon end because the paper was being sold. He would be acting chief executive 'for the early month or two' and expected to be involved in company affairs seven days a week. Dick, his deputy chairman, chipped in that he would be able to allot '75 per cent of a conventional week' to the company's affairs.

But the matter still bothered IBA officials. Lady Plowden raised it again later in the hearing. Wasn't there a danger that the proposed new chief executive might try to 'muscle in' on the presenters and prevent their having the influence they sought on programme philosophy? Peter replied that as executive chairman he would be careful to appoint a chief executive with an eye to compatibility. 'I have no other job. I'm determined to make it a success. I'm absolutely clear we need a chief executive.'

Members of the Authority had been allotted in advance the questions they had to put to the applicants. It fell to Lady Anglesey to raise the potentially tendentious issue of relations with ITN. Here Peter was forthright. 'There's not a shadow of doubt in my mind that anyone who thinks they can create a primary news organisation with world-wide stringers [local correspondents] and so on, on the financial base of breakfast television, is deceiving himself. Absurd duplication. ITN is a superb news-gathering service. We wish to work closely with them and co-operate. . . . ITN is a news service and wishes to ensure the editorial integrity of its material. We are a programme company and want to ensure the editorial integrity of our programmes. . . . There are no problems here which experienced, practical people who want to work together can't work out. But it can't be worked out in five minutes.'

Lady Anglesey persisted. 'What happens if it can't be worked out?'

Here Peter changed tack. 'I'm not saying it can't be done without ITN. It's the best way of doing it but we can do it ourselves if we have to.' (But hadn't he just said he'd be deceiving himself to think TV-am could do it on its own? He had, but Lady Anglesey's turn was over and the subject was not raised again.)

The IBA had already decided amongst themselves that the new breakfast station should not begin broadcasting until 1983, to allow time for Channel Four to get established following its proposed launch in November 1982. Like the other applicants, TV-am had based their plans on a start date of January 1982. Peter was asked whether their proposition would still be viable if the start were to be delayed. He replied that obviously they could not hang about until, say, 1985, but they might be able to wait a few months beyond January 1982. 'We have to earn a living between now and then,' he pointed out, but although a delay would be expensive in terms of cash flow it might be an advantage technically. 'The iron is hot,' he declaimed, then modestly added 'to use a cliché. People are raring to go' (to use another cliché). But whatever the Authority decided 'we must and will happily abide by'.

The financial prospects were discussed only briefly. Michael Rosenberg had taken with him some computer printouts of revenue expectations in relation to certain variables but, like the lady who took her harp to a party, nobody asked him. Peter said TV-am were projecting an annual revenue of £15 million once they were into their stride. That related to the total revenue of the commercial TV network of £500 million. Only a third of TV-am's revenue, he thought, would be forfeited by the existing ITV companies. Mrs Coulson for the IBA said she noted that TV-am's revenue estimate was higher than that of the other applicants; even so she thought it might be conservative. Optimism was not confined to the applicants.

The question of the studio, too, was raised only briefly. 'We're proud of it,' said Peter, referring to Ewart's facility in Wandsworth. The ingenious studio design was 'lean and cost effective'. Nobody inquired whether the deal with Ewart was watertight. Then Nick Elliott and Michael Deakin spoke of the actual product. 'Programmes are living things,' Deakin declared. 'They can work only if they seem to make everyone happy.' He told the Authority members of the time he had spent on the production staff of Jack de Manio's old morning radio show. 'We'd pattern ourselves like a living animal', he confided mysteriously.

Dr Morris of Glasgow, the house wit, asked a question that seemed to have been inspired by the cyclist who had spoken out at Darlington. Maybe he felt obliged to articulate the northern view, for his question was how children could be prevented from watching the programme and missing school as a result. Anna, who had some expertise in children's TV, said some of the material would be partly educational, so the schools might want to use it. Peter, thinking as ever on his feet, was struck by what must have seemed a splendid idea at the time. He said the presenters could urge children to go to school, and remind them not to forget their gym shoes and dinner money. It would, he affirmed, be 'an important social contribution'. If any of the Authority members thought this a bit over the top, nobody said so. They quickly went back to discussing the mission to explain.

The session, already stretching well beyond the allotted hour, was nearing its end when Peter, as team captain, suddenly realised that one key member was not pulling his weight. He observed that David Frost had not said very much. David needed no stronger cue to launch into what was evidently planned as a rousing peroration: 'I think we've shown how excited we are by this new frontier. We think we can convey this excitement to the audience, not least by the chemical reaction between people' – a thought he was later to develop as the notorious 'sexual chemistry'. He went on: 'There could be a new intimacy in the morning with the audience. If it's good enough it will stimulate conversation and enrich people's day.'

It was an upbeat ending to what everyone in the room could now sense had been a highly successful interview. They had all given of their best. John Birt's advice had been taken on board. They left and walked along Knightsbridge, where a large table had been reserved for lunch at the Hyde Park Hotel. If the early part of the TV-am saga was measured by airborne meetings, now it seemed a succession of merry parties.

Before they sat down Dick Marsh drew Peter aside. There had always been a certain stiffness in relations between the two, perhaps based on politics – Dick being a defector from Labour and Peter still a loyal party member. Jonathan Aitken was later to ask Peter, jokingly, whether he had ever stolen a girlfriend from Dick or delivered some other comparable slight. Peter protested that he certainly had not; but he too could sense the tension and was therefore gratified by what Dick had to say to him that lunchtime.

'Look,' he said. 'You're the chairman and I'm the deputy chairman. Let's not waste time. If at any time you feel it's not working or I'm getting under your feet and you don't want me around, say so and I'll go without any fuss or bother.' The statement touched Peter. He thought it fine and honourable and he told Dick so.

'I don't want you to go,' he assured him. 'You've got a big contribution to make.' Peter admires straightforwardness beyond most other qualities. That single conversation convinced him for some time afterwards that he could rely on Dick.

CHAPTER 3

Experiments in Chemistry

It is hard to imagine a less convenient day to make a vital and long-anticipated announcement, demanding the presence of some fifty senior figures in public life, than the Sunday between Christmas and the New Year. That was the day the IBA chose to reveal who had won the franchises not just for breakfast TV but for the fourteen regions. And they insisted, so far as they could, that the chairmen of all the rival groups should be on hand to learn whether their bids had succeeded, careless of the disruption to their seasonal holiday plans.

Lord Lever, for one, felt he might have better things to do with his Sunday than hang around the IBA's offices to take part in what he felt sure would be a charade. So he telephoned Sir Brian Young, director-general of the Authority, to ask whether he might be excused. Could he not sit at home in comfort and have the news relayed by telephone? No, Sir Brian replied. That would not do at all. The Authority were anxious, for all kinds of reasons to do with their dignity and prestige, that as many as possible of the group leaders should turn up. Maybe they had seen too many programmes like the annual Oscar ceremony from Hollywood, where there is a dreadful sense of letdown if any of the winners is not there to receive the award in person. Lord Lever sighed and agreed to go along, more for the sake of his associates than for himself. But he did wonder idly, with a small flutter of excitement, whether Sir Brian's insistence that he should be there was a hint that his group was going to win the breakfast contract. If that was so, the sacrifice would be worth it.

A charade it certainly was. Since quite early in the morning, relays of applicants for the regional awards had been arriving at IBA headquarters. Each was taken to a separate office by a separate usher and fetched a cup of tea. After a time another usher came in with a sealed envelope. Trembling, the victims would open it and were maybe expected to squeal with joy or cover their eyes with pain. The victors went in to shake hands with Sir Brian Young and Lady Plowden and were asked to go back for a press conference at 4 p.m. The losers were free to leave, presumably heading for the Serpentine in Hyde Park, where they could tear Lady Plowden's rejection letter into little pieces and feed it to the ducks. 'Psychopathic bureaucracy,' was how one indignant loser described the procedure. 'A fiesta of conceit.'

The breakfast bidders had been summoned to play their role in this melodrama at 2 p.m. David thought this a good pretext for yet another party, where the gang could hold hands, cry on each other's shoulders, celebrate with champagne if that was going to be appropriate, and generally rediscover what good friends they had become. All the principals turned up except Angela, at home in the West Country, and Robert Kee, visiting Washington. Jonathan Aitken dropped in on his way to learning that his Yorkshire consortium had been rejected. Even Keith Ewart, the studio man, had been invited, but he was on his way to Los Angeles for a holiday. 'Have a glass for your absent friend,' he wrote to David.

Shortly before two, Peter left David's house for the IBA. He could easily have walked – it is scarcely more than five minutes – but instead Michael Rosenberg took him in his BMW. Michael was singled out as chauffeur because he had a phone in the car. Anxious to get word to his associates as quickly as possible, Peter was unsure whether the IBA's mania for pomp and security might have led them to lock all the phones while their decisions were made known. If he could not phone from inside he planned to race out and get word to Michael, who would then phone back to David's house from his car.

While Michael waited across the road outside Harrods, Peter went into the IBA to claim his envelope. It contained a three-page letter from Lady Plowden announcing that his group had been awarded the franchise but would not be allowed to start broadcasting until 1983, because of the IBA's desire to see Channel Four launched first. The exact date would be settled after discussion between the Authority and TV-am. 'We realise the strength of your desire to start . . . as soon as possible but we gained the firm impression from what you said at the interview that your group would certainly stay together until 1983.' There followed a touch of the big stick, when Lady Plowden wrote that if the delay caused serious problems the IBA would want to know as soon as possible, as other groups would be able to provide the service. She also stressed that she would like to announce at that day's press conference the names of the hitherto secret members of the team. The letter concluded: 'May I send you and your colleagues all my personal good wishes for what will be an enormously challenging venture? I am sure our confidence in you all is not misplaced and you will provide a most worthwhile addition to British television.'

Peter had scarcely been so excited since he was made head of house at Winchester. There he had received the accolade from the headmaster. Here he was asked to go and see Sir Brian Young, who was as delighted as Peter, for he had been the keenest supporter of TV-am amongst the professional staff. Sir Brian said it was important to have immediate confirmation that the delay in starting the new station would be acceptable to all members of the group. That gave Peter the excuse he needed to use the phone, which had not been locked after all.

David took the call while the others formed a tight, tense semi-circle round him. Peter explained that he was sitting with the Director-General, and David took that as a plea for a certain amount of decorum. Then Peter announced as dispassionately as he could that they had won the contract.

'I see,' said David calmly. 'And have we won it outright or do we share it with anyone?' (There had been speculation that the IBA might force a merger of two of the competing groups, or give it to one of them jointly with ITN.) Peter told him.

'I see,' David said again, in a flat tone that belied his racing pulse. 'We've won it outright.' At that the gang around him burst into spontaneous cheering. David quickly placed the mouthpiece to his chest in the hope of muffling the sound at the other end. It was useless. The roar was easily heard by Sir Brian, who grinned broadly.

Before going back to the IBA for the press conference, Peter had time for a brief return visit to David's party. Another enormous cheer rang out as he entered the room, like a batsman going back into the dressing room after scoring a century. Flushed and delighted, he unbuckled his metaphorical pads and helped David open the champagne. Press photographers, arriving already, were invited in to take celebratory pictures.

Esther Rantzen found herself talking to Nick Elliott. 'Wow!' she exclaimed. 'So it's for real. Are you for real?' Nick was taken aback by the question. 'I'm for real,' he replied. 'Are you?'

Esther assured him she was, but he wondered why she had asked. Was she starting to have doubts about her participation? By the following Christmas, both had left the gang.

The IBA's decision to award the contract to Peter's group was a close one, reached after long and sometimes rancorous discussion. The two senior officials on the television side, Colin Shaw and David Glencross, advised strongly against TV-am. Their experience in television made them sceptical about whether Peter's fine words could be translated into practical action. They raised the spectre of London Weekend, which had been given its franchise in 1967 on a similarly well-intentioned set of proposals (Brecht against Match of the Day, as the industry joke had it) and had all but foundered, being forced to abandon much of their highbrow prospectus in order to survive. They thought Peter was woefully lacking in the experience needed to head a TV company. As for the star presenters, for all their protestations about how keen they were to work together, they were in essence a collection of prima donnas, unlikely in the event to blend into an effective team.

Shaw and Glencross were much more impressed with Christopher Chataway's group, and in particular with Harold Evans, whose performance in the interview had been a *tour de force*. Evans, they felt, had a better understanding than Peter Jay of the mechanics of collecting news and putting it across. He was more down to earth, with working-class origins that gave him an instinctive grasp of what people wanted to watch, a skill for which Peter's Winchester and Christ Church education had scarcely trained him. They were impressed, too, with the quality and experience of the technical people in the

Chataway group – especially Mike Wooller, the head of documentaries at Thames, and John Fairley, former head of news and current affairs at Yorkshire, where he worked closely with Michael Deakin. The pity was that these two had not been allowed to say much at the hearing. Evans had tended to speak for all of them, giving the unfortunate impression that it was going to be pretty much a one-man band. John Birt had cured Peter of that fault at rehearsal.

There was some support for Shaw's view amongst the lay members of the Authority. Tony Christopher of the Inland Revenue Staff Federation expressed his reservations about Jay quite vehemently. Lady Plowden herself, although she liked Peter's style, on balance preferred Evans's brisker approach, his chirpy confidence. Amongst the officials, the strongest backing for TV-am came from Sir Brian Young and his deputy, Tony Pragnell. Sir Brian thought Peter, with his lucid intellect, would make a welcome adornment to independent broadcasting. He was attracted by the mission to explain and thought it could be made to work. So did Mary Warnock, Peter's tutor at Oxford. She thought him a brilliant student and saw no reason why he should not become an equally brilliant organiser of a television company. By the time the decision came to be made, it was down to these two groups – Jay's and Chataway's. At the final poll there were just two votes in it.

History is full of if-onlys, but it is intriguing to speculate what would have happened had the vote gone the other way. Evans would have quit journalism for television and could not have been appointed editor of *The Times* by its new owner, Rupert Murdoch. So he would have avoided his public humiliation by Murdoch a year later. Peter Jay would not have become a television tycoon and would not in turn have been humiliated by his dismissal in 1983. Two notorious acts of blood-letting in the media would thus have been averted.

The day after the franchise award was made, the BBC announced that they too would start a breakfast programme and it would be on the air by the spring of 1982. This came as a surprise to the IBA and to many in the TV-am consortium. The BBC had spent much of 1980 pleading poverty, agitating for an increase in the licence fee that is their primary source of revenue. They gave the clear and deliberate impression that they could scarcely afford to fund their current level of broadcasting, let alone take on an additional three hours a day. And there were blank hours in the afternoon which might have been thought to rank higher in priority than the early morning.

Yet that assumption about the BBC's intention failed to take into account the Corporation's fiercely competitive nature. The BBC mandarins resented the creation of independent television in the 1950s and even now, nearly 30 years later, that attitude persisted, coming to the fore in a series of acrimonious disputes about the coverage of sport. For several years the BBC used to insist on duplicating ITV's coverage of the Derby, under an agreement that gave both channels equal access to major national events. ITV had the contract for the Epsom meeting as a whole, yet the BBC would trundle all their gear down to Epsom just for that one race. There had been similar absurdities in televising football. The BBC's attitude seemed to be based on the paranoid fear that they would be crushed under the powerful independent network unless they sturdily stood up for their rights. Sue Summers, TV writer for the *Standard*, called it 'a fierce combative lust to compete with their opposition at any cost.'

There would have been nothing to stop the BBC starting breakfast television many years ago had they thought it desirable. They were deterred by the same considerations about union overtime rates as had inhibited the independent companies. Yet those agreements could have been renegotiated – as they were when the plans were later finalised. The reason they had not done it before was that they believed there was little demand for it. And when they decided to go ahead it was not because there was sudden evidence that altered that belief but because they did not want to let ITV get away with it. BBC people spoke of the carry-over effect. Such was the lethargy of many viewers, one school of thought had it, that they were reluctant to change the channels of their set. If they switched off at 9 a.m. tuned to ITV, they would still be tuned to it when they started viewing again in the evening, and they would stay there. (The argument cannot be applied to modern electronic sets. They switch themselves on to BBC-1 no matter what channel was being watched before.) 'We felt,' said a senior BBC executive, 'that we just had to be in the game.'

The BBC is not a commercial organisation. Its revenues are not at all dependent on the number of viewers it attracts. People have to pay the licence fee even if they only watch the independent channels. Yet that seems to spur the BBC to be more competitive, not less. Once they lose that impetus, they fear they will sink into elitism or lazy mediocrity.

At the BBC Television Centre, then, there was little doubt that a response to the ITV breakfast initiative would have to be made. Just what kind of response was a subject of intense debate. One idea was for something dubbed 'radiovision'. This would have involved adapting the highly successful 'Today' show on Radio 4 to make it into a joint radio and television broadcast. (Something like that was tried in the 'Election Call' programmes on both radio and television during the 1983 election.) Aubrey Singer, then controller of BBC radio, was in favour. He thought there was a logical unity in the concept. People would wake up to the radio, take a portable set into the bathroom, watch TV during breakfast, and carry on listening while driving to work. Singer appointed a committee to look into it and their report was favourable.

But there was vociferous opposition out at the Television Centre, where the idea of sharing any new programme with radio, the poor relation, was an anathema. It was felt that the plan would so alter the character of a successful radio show as to destroy it − and would make for dismal television into the bargain. So they set up their own committee which, to nobody's great surprise, concluded that any breakfast programme should be pure television, not some misbegotten hybrid. At the beginning of 1982 Singer moved to take charge of the Television Centre and soon changed his opinion. Radiovision was ruled out. As Singer explained to colleagues: 'Responsibility sharpens the mind.'

When the New Year festivities were over, Peter and David created a steering committee to oversee the first stages of forming the company. It consisted of those two plus Dick Marsh,

Michael Deakin, Michael Rosenberg and Nick Elliott. Their inaugural meeting was held in David's drawing room in early January 1981. The first thing they decided was that they wanted Peter to be chief executive. Their motive was partly financial. They were two years away from the time when, according to the IBA's schedule, they would go on the air – although Peter was already determined to try to get the start date brought forward. It would in any event be much more than a year before they began earning any revenue. Why carry the burden of a highly-paid chief executive for all that time? And hadn't Lord Thomson been broadly hinting at the hearing that it would be best if Peter did both jobs? It would make far more sense to wait until the on-air date approached before going out to look for a new man.

Peter was pleased by the suggestion. He had no doubt whatever that he was capable of running the company successfully, just as he had not doubted his ability to win the franchise for them. Who knew, he might prove such a brilliant success that, come 1983, the board would feel they did not need anyone else. Yet there was one dissenting voice. Dick Marsh pointed out that they were proposing to appoint as chief executive someone who had never held any executive role in his life. Intelligent, yes; articulate, certainly. But he had never managed anything (unless you counted the Washington embassy, and Dick did not). And he was going to have to oversee the complex business of hiring architects, buying in sophisticated equipment, recruiting staff. Dick foresaw danger. But he did not press the point and the decision to give Peter the dual role was unanimous.

The most urgent priority for the team was to finalise the deal with Keith Ewart over the studio. One of David's first actions, on learning of the award, was to have a message sent to Keith on his transatlantic flight. It was a characteristic gesture and Keith appreciated it, but during the long flight he had time to ponder what it would mean for him. Things would certainly never be the same at his studio again. Ewart was then fifty-four, a twinkling and utterly charming man who on first acquaintance appears deceptively unworldly. His conversation is a whirl of bubbling enthusiasm. Once a musician in Kenya, he returned to

Britain and became a professional photographer. By the early 1950s he was doing work for the advertising agency J. Walter Thompson, just branching into commercials with the advent of independent TV. He and the agency's creative people learned the art together and for eight years he was making nearly one commercial a day. With the money he made he brought an island site in the middle of Wandsworth's formidable one-way system and opened there in 1967 a television facilities studio, for hire to makers of commercials and independent producers.

Thus it was his own place, built with the fruits of a lifetime's hard work. With such as David, Peter and the famous presenters doing their dazzling thing, it was unlikely he would ever again be able to call it truly his own. Still, he would soon be approaching sixty, and would have to think of pulling out anyway. And they seemed warm, understanding people: he was sure they would treat his staff right. He had built up a team he regarded as the most talented in the business and took trouble to make them his personal friends. He did not want them swept aside in the crush of new people. Yet although Keith could sense that there might be pitfalls along the way, he had no doubt, as he stretched back in his aircraft seat, that accommodations could be made.

Peter and David had no doubts either. On the day after they were awarded the franchise they repaired to Wandsworth to be interviewed for Thames Television. As the commentator said in his introduction, it was not a bad part of town, but hardly the place you would expect to find some of television's best-known and best-paid stars. Standing outside the deserted studio entrance (it was closed for the Christmas holiday) Peter and David were their usual effervescent selves. It was, said David, 'the perfect location – so close to the centre of town'. Peter agreed. It was 'absolutely ideal', he said, explaining that there was even vacant land adjoining the studio where offices could be built. Moreover, from the point of view of the politicians they would be interviewing every morning, it was 'the closest television centre to Westminster'. This was patently untrue. London Weekend, near Waterloo, is very much closer. Peter must have been a victim of the disorientation that, as every South Londoner knows, grips strangers beyond Clapham Junction.

The segment ended with the interviewer reporting breathlessly that residents were already speculating what Angela and Anna would look like at five in the morning. In the end, speculation was all they were left with.

Keith was upset when he returned to Wandsworth on 8 January and learned of the interview. He thought it cavalier of David and Peter to assume proprietorial rights over the place when scarcely any details of the arrangement had been discussed, much less agreed. But he was prepared to put that down to high spirits. He wrote to David thanking him for his in-flight message:

> I am glad the best men won. In the good old days (come back Lord Reith) the BBC was an example to the world. I honestly believe that now it is *your* turn. I hope people come from far and wide to see your station for its spirit, its programme content and its technical standards. . . . Perhaps soon we should discuss broadly how now to proceed.

Keith's letters tended to be discursive and informal. He thought business could not be properly conducted unless warm human relations had first been established. It took two more such letters – one to Peter and one to Dick Marsh – before he was able to set up a meeting with the steering committee over champagne at David's house at the end of January. That was when difficulties began to become apparent. Keith raised the questions of the effect of the proposed new construction on his existing studios, of relations between his engineers and those of TV-am and of compensation for loss of his revenue during the building period. Dick felt it necessary to write after that meeting to assure Keith that 'we genuinely wish to work closely with you'. He looked forward to the day when the discussions could be translated into 'the soulless prose of regrettably essential formal agreements'.

Keith attended the February meetings at TV-am's temporary Mayfair offices to select an architect for the work. The man chosen was Terry Farrell, known as an innovator in the post-modern style. He said the necessary alterations and new construction could be completed for £2½ million. Wandsworth Borough Council were formally approached for planning permission. Yet still no progress was being made on the points that worried Keith. They agreed to have a formal meeting on 27 February at his solicitor's office in Gray's Inn, where lawyers and accountants could thrash out the difficulties. Keith did not want to attend the whole meeting – that is what lawyers and accountants are for – but he did want to go along and say a few words at the beginning. Their effect was startling. A few days earlier Prince Charles and Lady Diana Spencer had announced their engagement and he began his speech by using that as an analogy with the proposed marriage between TV-am and his studio.

Lady Diana has advantages over me other than the obvious ones. Firstly she has no doubt as to the ability and willingness of her future husband's family to keep her in a style which is at least adequate. And secondly her freedom of action was not prejudiced by her Prince announcing the engagement from outside her modest dwelling while she was away for Christmas and before she had said yes.

He was adamant that whatever agreement was reached the land must continue to belong to him. 'In this case there is no dowry.' He insisted on protection for 'my dependants and, more important, my old retainers'. He thought they were nearing a point when the facilities left to Ewart's in Farrell's scheme were 'below the minimum acceptable'.

We will have to close down our operation progressively between now and August, and then totally. We will have to move all our equipment and to reinstall it in new areas. God knows how long it will take us to get the bugs out of it again. God knows if we will get many of our old clients back after one year plus shutdown. Our image if not our character will change. We are being asked to put ourselves at risk at the very moment that we could fairly expect to enjoy, through IBA 4 and satellites and the television explosion generally, the fruits of our patience over these lean years.

81

He therefore insisted that those losses be fully covered by TV-am. That would, it was later calculated, cost something close to £1½ million. But it was not just a question of money. What was surely the key part of his speech came a few moments later:

> The fact that I am not prepared to lose the apartment in which I daily practise my clarinet may seem a trivial reason for holding up the building of your national network centre. But to me that flat, located at my place of business, represents the reward for a lifetime's hard work. Our agreement on paper is worth less than our mutual understanding of each other's very different aims, objectives and even values. I must confess that the TV-am scheme is by far greater than I and possibly any of us had anticipated. Maybe it brings into question anew the practicality on both sides of putting it on our site.

Just when it seemed that he was going to bring down the curtain there and then, Keith concluded on an upbeat note: 'I am sure that with Terry Farrell's genius Ewart's problems are surmountable. From TV-am's point of view, the advantages of Ewart's are clear and considerable. . . . The alternative to all of our unequivocal acceptance of this scheme would surely mean a delay of some months to the first airdate. During these months a cheaper answer might be found but it could not be a better one.' With that Keith left the meeting and headed back to Wandsworth, leaving a room full of flabbergasted negotiators to work something out in the spirit of that statement, if they could agree what that spirit was.

As soon as he arrived back at the studio Keith had a telephone call from his accountant at the Gray's Inn meeting. The TV-am people, he was told, were insisting on a 125-year lease on the land where the office complex would be built. He would not think of it. He instructed the accountant to end the meeting forthwith.

The steering committee now had two options. They could persevere with Keith, accommodating his desire for sovereignty over his premises and working towards agreement on a compensation figure. Or they could find a different site and build an entire new studio from scratch. It did not take them long to persuade themselves that the glamorous second course, giving them a place of their very own, was the one to adopt. They circulated the planning departments of the Inner London boroughs and asked them to suggest suitable sites. Camden Council steered them towards a disused garage and motor repair centre in Hawley Crescent, backing on to Camden Lock. The cost would be a great deal more than moving in on Ewart's but, now they had won the franchise, they anticipated no difficulty in raising extra capital.

When this was discussed by the board, one director wondered whether, before they committed themselves to this expenditure, one more attempt to come to terms with Keith Ewart might not be made. Jonathan Aitken had paid a brief visit to Wandsworth soon after the Gray's Inn debacle, looked round the studio and told Keith: 'If I can be a mediator in this, let me know.' A few days later he invited Keith to lunch at the House of Commons, on 17 March. At lunch Jonathan made a surprising proposition. He said he had, in Aitken Telecommunications, a media company quite separate from TV-am. On behalf of that company he would like to make an offer for the studio. He offered £2½ million for a two-thirds share in the business. Keith turned him down, but did agree to try to resume negotiations with Peter.

It was too late. That very evening Keith received a phone call from a reporter from *Broadcast* magazine, quoting a statement from TV-am that, since he had failed to respond to their proposals, they were looking at other sites. The statement mentioned the difficulty of communication 'on such an emotional level'. Keith dashed off a pained letter to Peter. He said he had been waiting for fresh proposals from TV-am. He would still not consider a 125-year lease but might now think of selling the whole facility to TV-am. He had turned down Jonathan's bid partly because it was for only two-thirds of the company. He certainly did not want to be a minority shareholder in what he would inevitably still regard as his own outfit. And if he did sell, it would only be if every one of his employees was convinced he would have a better future with TV-am.

'Now, Sir, I can hardly be less emotional than that,' he concluded. 'The funny thing is I still quite like you all.'

83

In this case, friendship was not enough. The Camden Lock decision had been made. Farrell now had to draw up new plans for the studio, using as much of the old garage structure as he could. The budget for this (the building without the technical equipment) was a bit below £5 million. By October, the final shape of the building had been agreed by TV-am, Farrell and the building contractors, Wiltshiers. Soon after that, work was under way.

In the spring of 1981 Esther Rantzen, already the mother of a young child, discovered she was pregnant again. The baby would be born in October and would be less than two when TV-am went on the air. Could she cope with getting up before dawn and making for the studios with two young children in the house? She thought not, and engaged in a series of earnest converations on the question with Peter. They tried to devise a work schedule that would avoid the early morning vigil but Esther felt it would be wrong to claim privileges over the other five. Reluctantly, she told Peter she would have to leave the gang.

She was unhappy, though not because she was particularly enamoured of the idea of breakfast TV. She was not one of those in the industry aching to explore what David had called the new frontier. Some of her colleagues did think like that. Ron Neil, part of her production team on 'That's Life', had told her how much he envied her the opportunity. Mike Townson of Thames, managing director designate in the consortium that included David Dimbleby, was another convinced he knew just what the public wanted on the screen to complement the crackle of the cornflakes. Esther had no such commitment to the morning time slot. Indeed, she had never really reconciled herself to the idea of quitting 'That's Life'. But she had, like the rest of the gang, relished the spirit of *camaraderie* that had grown between them. She felt bad about letting the others down. And in material terms, she thought it might be an act of the most reckless irresponsibility to forfeit her piece of the equity, her insurance policy for when the puppet lost her magic.

As far as Peter and the others were concerned, it could scarcely be said that the loss of one of six presenters was a mortal blow,

A confident start: as work begins at Camden Lock, Peter Jay proudly shows off Terry
Farrell's model of the TV-am building.

Top left: Esther Rantzen, the first defector from the original gang. *Top right:* Lord Marsh, deputy chairman of TV-am and Peter Jay's earliest critic. *Bottom left:* Greg Dyke, who gave TV-am a new look after arriving as editor-in-chief in May, 1983. *Bottom right:* Lord Camoys who had to ask Peter for his resignation.

Above: Michael Deakin, director of programmes, at Camden Lock during the early stages of building. *Below:* The competition: Frank Bough and Selina Scott of BBC TV's Breakfast Time.

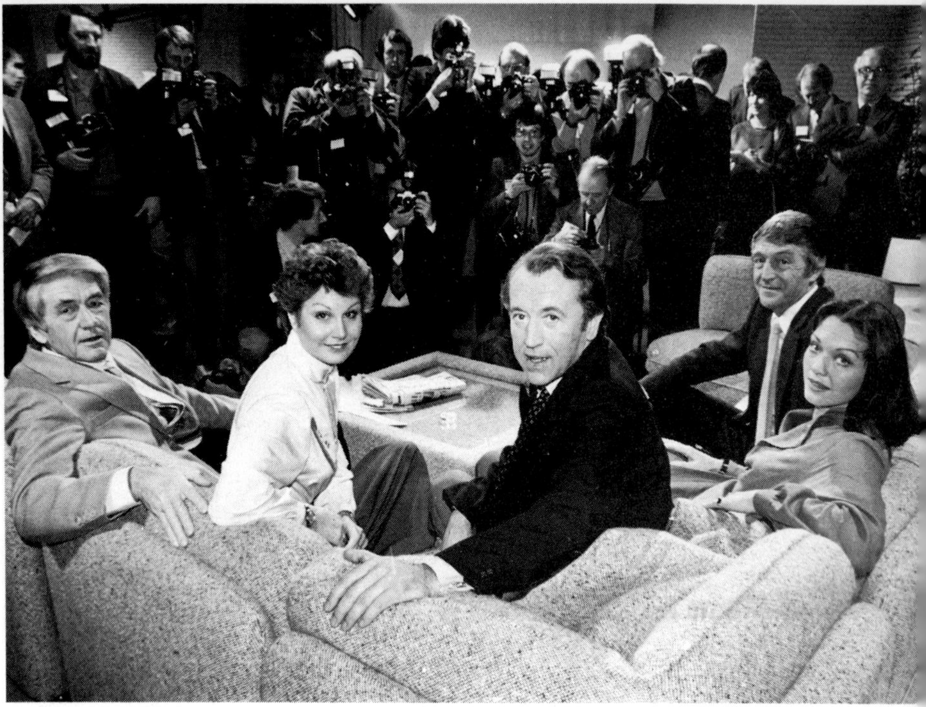

Above: The Famous Five. *Below:* Something to celebrate: Lord Thomson, chairman of the IBA, with Angela, Michael, and Peter, in Peter's office on the first morning of transmission.

bove: Egg-cups and all. The exterior of the TV-am building. *Below:* Crowds inside the uilding on the opening day.

Above: Anna and Angela leading the 'Jay must stay' demonstration on 18 March, 1983. They did not know that he had already agreed to resign.

Below: The day after Peter's resignation, David marries Lady Carina Fitzalan-Howard at Chelsea Register Office.

Right: A month later, Anna talks to reporters outside her house after being fired.
Below: Happier days, at the party to celebrate TV-am's winning the franchise for breakfast television. The three on the right are Michael Rosenberg, Jennie Bland and Nick Elliot.

The Aitken cousins, victors in the boardroom battle. *Top:* The sauve Jonathan, who served as chief executive for the month following Peter's departure. *Bottom:* Timothy, the rough diamond who took over from Jonathan.

although Esther's record suggested that she was the most innovative of all of them in production terms, with the sharpest instinct for popular taste. Far graver was the departure some months later of Nick Elliott, their designated director of programmes. Nick had stayed with London Weekend in the months following the award of the franchise, in part because he thought it wrong to burden the new company's finances with his salary while they were still far from earning revenue. He was working on the plans for TV-am in his spare time. He had been genuinely excited by the prospects for breakfast TV but part of his reason for agreeing to join had been negative. There was a log-jam at London Weekend of executives just below the top tier. It happens in all organisations from time to time, especially in the media: too many promising young people jostling for too few senior posts.

Nick had been head of features for five years and was feeling frustrated when Peter first approached him in 1980. There was no realistic sign that any of the jobs he coveted in the company would soon fall vacant. He was fond of London Weekend. It was a civilised, companiable and well-run concern. It was close to his South London home. But he feared that it was starting to get altogether too comfortable, that he was in danger of losing his professional bite. It was like sinking into a cosy armchair and getting more and more reluctant to climb out of it as the cushions settled themselves ever more snugly around him. Peter's offer had come just in time to tip him from the chair before he dozed off.

In September 1981 John Birt, his immediate superior as controller of features at London Weekend, told him that Michael Grade, the director of programmes, was going to leave and that he, John, was going to succeed him. That set Nick thinking. Now at long last the logs were getting unjammed. That one move, he could see, would set off a whole series of chain reactions. It would be the start of a new era in the company and suddenly he felt pangs of regret at not being part of it.

A few weeks later John Birt told him that, if he were ever to change his mind about TV-am, the post of controller of drama and arts programmes at London Weekend could be his for the asking. It was a tremendous temptation, a new post

85

combining elements of two old ones, representing much of what interested Nick in television. He felt he may have had enough of current affairs. So despite his commitment to TV-am, he did not dismiss John's suggestion. The more he thought about it the more attractive it became. After keeping it to himself for a week or more, he began a series of discussions with Peter and David.

Quite apart from the London Weekend offer, Nick was beginning to have reservations about his role at TV-am. His relationship with Michael Deakin remained ambiguous and it began to matter more as the time neared for detailed planning of the new programmes. Early on, John Birt had warned him how vital it was to have clear lines of command. There were decidedly none of these in relation to Michael, who had accepted Nick's *primus inter pares* status with reluctance at a time when nobody was really convinced that they would win the franchise. Was Nick swapping his comfortable armchair on the South Bank for a bed of nails in Camden Town?

Thinking further, he could not see TV-am as a job for life. They would go on air, it would be a success and he would enjoy a sense of achievement at what he had created. But what then? All he could do was keep on going at that level, and he could foresee the challenge of that soon wearing thin. TV-am was a one-product organisation: no chance of a job as head of drama because there is no drama. He reckoned that after a year he would be looking round for something else. By contrast, the position he had been offered at London Weekend was nothing if not long-term. Sometimes it took years for major drama projects to get from the drawing board to the screen. Despite David and Peter's powerful urgings, and despite the tugs of group loyalty, Nick decided to stay with London Weekend.

There was never any serious thought of replacing him. Michael Deakin had always coveted the post of director of programmes and was keen to have it. His former role as director of features was one that had been created specifically to accommodate him and did not now need to be filled. Whatever doubts Peter and the others had about him, he was a founder member of the company and to bring in somebody new of equal status would be seen as a slight, even an act of betrayal. Yet he had never had a job with as much responsibility before and, more important, he had little experience in making programmes live or in the studio. He was essentially a film-maker. He would tell people before Nick's departure: 'I'm not really here for the daily grind. I'm here to provide the sparkle.' It is an open question whether the IBA would have awarded TV-am the contract if they knew a man with his particular qualifications would be director of programmes. But now the contract was won the IBA agreed to his appointment without demur. Michael was as aware of his deficiencies as anyone but cheerily told himself and the others that all he needed was a strong editor to support him, someone for the daily grind.

Yet the man he had in mind for the post did not fit that bill at all. Hilary Lawson was a twenty-nine-year-old producer who had worked with both Michael Deakin and David Frost at Yorkshire TV, where he had been employed for six years. A lean, bearded Oxford graduate, he was certainly bright but crucially did not have the basic experience of producing daily current affairs television — the area where Michael was also weakest. Though less scintillating than Michael, Hilary was similar in background and in his range of cultural interests. He was a characteristic Deakin appointment. Michael had little time for the dour competent technocrats who play vital roles in the mechanics of making programmes, as their equivalents do in newspapers. He preferred young, unmoulded people of glittering promise. Hilary indisputably came into that category.

Michael did try to strengthen the editorial team but without success. Early in 1982 he approached Greg Dyke, a promising young editor at London Weekend, about to launch the highly rated 'Six O'Clock Show' on Friday evenings. He wanted Dyke to be joint editor at TV-am with Hilary, both subordinate to Michael. 'It will never work,' said Dyke, who thought joint editorship a recipe for disaster: nobody would know who was in charge.

During 1981 Michael and Peter began having differences about the programmes and about whose responsibility they finally were. Peter thought, as an experienced journalist and a joint creator of the mission to explain, that he should naturally play a role in transforming the idea into reality. Michael disagreed. He complained that Peter was acting as though he were the editor of *The Times*, the only one steering. Peter, for his part, thought *Michael* was the man with delusions about editing *The Times* and was portraying *Peter* as the interfering proprietor, going beyond the role custom found acceptable – the Rupert Murdoch to Michael's Harold Evans. That was not how Peter saw himself, but when Nick Elliott came down in support of Michael he conceded the point gracefully, for he was soon so involved in administrative details concerning the new building and recruitment that he had precious little time to do much about the programmes anyway. He limited his interference to continually urging Michael, quite in vain, to draw up plans, schedules and philosophical outlines of the shows he intended to make.

Since in the early days most of his colleagues were still working for their other companies, Peter thought he should get them away for a spring weekend to concentrate their minds on TV-am. He thought first of taking them to a seaside hotel but Jennie Bland offered the use of her substantial Victorian country house at Abbots Worthy, near Winchester. With nine guest bedrooms, plenty of reception rooms and a tennis court at the end of the extensive garden, it was an ideal venue. Angela Rippon could not go but except for her all the presenters were there, plus Michael Deakin and Nick Elliott, who had not yet decided to quit.

It was an invigorating weekend. The gang sat in groups firing ideas at each other and writing copious notes, breaking off only briefly for tennis and a guided tour of Winchester by Peter, glowing with fond schoolboy memories. The ideas covered all aspects of programming. Somebody suggested that George Thomas (now Viscount Tonypandy), about to retire as Speaker of the House of Commons, should do a Parliamentary commentary. Preliminary soundings had been taken and he seemed agreeable. Jennie Bland, the hostess, passed on news of something she had encountered in America, a tie-in between a TV station and a fast-food chain, where viewers are given a password which entitles them to a discount that morning at the chain's branches. Great, said the others.

Then there was the magic wand. This was a newly-invented device using technology similar to those blocks of broad and narrow black lines on packets in supermarkets. It would be used mainly in children's programmes. Using the wands, viewers would be able to answer multiple-choice questions on the screen and could tell if they had them right. The inventor had apparently been in touch with Michael Rosenberg, who was full of enthusiasm. The weekend was intellectually stimulating and hugely enjoyable.

Nick Elliott was deputed to take a note of all the ideas and blend them into a single written proposal. When Nick left, the task was transferred to Michael Deakin. But when almost the same group met again at Jennie's house many months later, they were dismayed to discover that nothing had been done to put the ideas into effect. Michael has a great aversion to making written plans. When, at a board meeting, Jennie Bland asked for fuller reports on programme proposals than the single uninformative paragraph he was wont to give, he snapped that he had better things to do than write long essays for directors to mull over. Peter was to come to think of Michael in terms Keynes used of a governor of the Federal Reserve Bank in the 1930s; not only had the governor no understanding of economic questions but 'he actively deprecates the exercise of the faculty of thought about these matters at all.'

At the root of Michael's procrastination was his reluctance to get to grips with Peter's mission to explain. While he was unsure what it really involved, of one thing he was certain: it was no kind of useful guide when it came to making programmes. He was not alone in that view. At a lunch of the founders at Barclay's Bank's city headquarters, Nick Elliott and Robert Kee were discussing the mission to explain. 'I've never understood what it meant,' Robert admitted. 'If it means good journalism I agree with it. If it's more than that I don't understand it.'

It may only have been a vague philosophy but it was to play a role — and not too helpful a role — in the negotiations that consumed much of Peter's time in the early months. One of the reasons the IBA preferred the TV-am application to the others was that it contained a firm and unambiguous commitment to working out a deal for news with ITN. Peter had reaffirmed this strongly at the hearing and in press interviews after he won the contract. Soon he sought talks with ITN to get the negotiations started.

It was never going to be easy. ITN had invested much work and many hopes in their own application for the franchise – first in persuading their shareholders, the programme companies, to go ahead with the bid and then in drawing up the document. John Freeman, Alastair Burnet and others had put their case eloquently at the hearing. It was clear that the verdict had gone against them not on merit but for political reasons, on some vague theory about the diversity of news sources. The ITN application had addressed that point directly. It maintained that diversity for its own sake was not necessarily a virtue, because the effect could be to weaken existing news sources.

Some six weeks into the new year, five members of the TV-am steering committee had lunch with five representatives of ITN at a Mayfair hotel. It was an amicable affair, replete with statements of high principle and willingness to co-operate. It was now clear that what Peter really wanted from ITN was a feed of raw news, rather than ready-made bulletins. The comparison he kept drawing, one he was to use often in the planning stage, was with the election night programme he had done for ITN in 1974. Robert Kee had been the anchor man and Peter had interviewed the politicians. He seemed to envisage that TV-am would be like that every day, with studio experts announcing and elaborating on the news as it happened. The snag with that idea is that on most days of the year not much happens at seven in the morning. Even an election programme falters when the results slow to a trickle.

In the meetings that followed it was apparent that agreement was going to be harder to reach than anyone thought. ITN is not a news agency and did not want to start being treated like one. Their product is news bulletins. That was all they were prepared to offer TV-am. Their price, because of the expense of keeping their operation going virtually round the clock, would be £6 million a year.

Peter was certainly not going to pay that for a service he did not want. He planned to abolish long bulletins of straight news in any case. Since his whole programme was going to be based on news he needed visual raw material to back up the factual agency reports the station would be getting anyway. He estimated the annual cost of gathering in that raw material, from sources other than ITN, to be £2 million. His case was that once the material was in the building it would be simple and inexpensive for TV-am to provide their own brief headline bulletins: after all, one thing they were not short of was newscasters. If it was important to ITN that they be credited for the bulletins he was reluctantly willing to concede that, although he thought it illogical. What he was not prepared to do was pay £6 million for it.

By summer, deadlock had been reached. Interviewed in *The Times* in July Peter said: 'If someone else is doing it [i.e. providing the news] then they are setting the agenda.' In the same paper the following month he told TV writer Elkan Allan: 'It's our ark of the covenant that the distinction between news and current affairs is what is wrong with televised journalism. So obviously we're not going to reimport a situation where we split our own programmes into news and current affairs. . . . We want viewers to understand why what's going on is going on, not just that it's happening. I personally watched television coverage of the Beirut crisis for three weeks before anybody told me that one side were Christians and the other Muslims and that had something to do with what they were fighting about . . . If ITN don't want to do it, we'll do it ourselves.'

It is scarcely surprising that ITN were not keen on making accommodations with someone who thus criticised their performance. David Nicholas, editor-in-chief of ITN, told Allan: 'We are a television newspaper, not a news agency. If TV-am want to comment on the news they are not going to mix up our reports with their interpretations.' He said he was worried that governments might exact reprisals against ITN reporters in foreign parts, for sins committed by TV-am's commentators.

It was not until late 1982 that an agreement was reached for access by TV-am to most of the material they wanted, on a pay-for-use basis through UPITN, the agency that sells ITN's material abroad. Peter maintained that this agreement gave him nearly everything he wanted. He theorised that ITN had delayed coming to an agreement in the vain hope that the IBA would bludgeon him into accepting the full service they wanted to provide.

Critics of the early programmes, however, said one of their chief faults, compared with the massively-resourced BBC effort, was in the poor quality of the news gathering and editing. They said Peter should have agreed to accept a fuller ITN service instead of expecting his brand-new station to provide an effective one right away.

That article by Elkan Allan was astute for other reasons. He calculated that the serious content of the proposed programmes was being overstated. He recalled how in America NBC's breakfast programme 'Today' had languished until they brought in J. Fred Muggs, a lovable gorilla. 'To capture the masses breakfast TV is likely to play the clown far more readily than is at the moment apparent.' He asked Peter whether he was planning to try something like the gorilla. 'I wouldn't rule out anything which entertained as it informed,' he replied cheerfully.

Allan's was not the only cautionary voice. In the *Financial Times* in the week after the franchises were announced, Chris Dunkley wrote: 'Don't be surprised if two years after that [1983] it is more Murdoch than Mirrorscope.' And in October David Hewson made the point in *The Times* that the famous presenters were 'irrevocably caught up in the image of British broadcasting as it was in 1980'. Was that, he wondered, what people would want to watch in 1983?

It was an imponderable question because, given Michael Deakin's reluctance to engage in too much forward planning, it was unclear exactly what people were going to be invited to watch. The presenters were left very much to their own devices as they went round the country promoting their new product to advertisers and to the broadcasting industry. They were given no brief about what to say, so invented things from the tops of their heads. David Frost works well in this fashion and it was he who helped launch a phrase that delighted headline writers but was to haunt them all in the ensuing months.

In April 1982 he was addressing a meeting of the Radio and Electrical Trades Association in Torquay. He told them, as he had told the IBA interviewers sixteen months earlier, that much of TV-am's appeal would lie in the interaction between the presenters, 'the chemistry between people'. Afterwards at a press conference Jack Bell, the television editor of the *Daily Mirror*, questioned him on the point. Would there normally be a man and a woman presenter? Yes, there would be. So the *sexual* chemistry would be important? 'The sexual chemistry is always important,' replied David. Another reporter elaborated the idea to suggest that the male and female presenters would be flirting with each other, and a legend – ultimately rather an embarrassing one – was born. If nothing else, it sounded more fun than the mission to explain, even if some questioned whether David might be a bit mature for it.

The IBA had already confirmed that it would be 1983 before the chemistry experiments would be allowed to begin. Peter had bombarded them with a series of elegantly argued letters explaining why a 1982 start date was essential, mainly for economic reasons and to get in ahead of the BBC. But in June 1981 the Authority announced the date as May 1983. Peter put his familiar bold face on this setback: 'We believed we put up a particularly strong argument for spring of next year . . . but we are not broken-hearted, alarmed or worried. Just very confident.' The following week Sir Ian Trethowan, the Director-General of the BBC, said he was now in no hurry to launch their service. There was no pressure to begin in 1982.

After still more persuasion from Peter, the IBA agreed to advance TV-am's launching date to February 1983. As it turned out, it was unlikely that they would have been able to meet the 1982 start date, seeing what a rush it was to get it ready for 1983. The equipment is the vital part of a TV studio and has to be ordered early. It is the subject of a running debate in the industry. With a host of new electronic cameras and gadgets being introduced almost by the week, many rival views are canvassed about what is best for a particular operation. One of the contentious issues concerns the width of the videotape needed for non-live material. One-inch tape has been usual in the past because the definition is so much better than on the narrower widths, but ¾" is more compact and less expensive to operate. Peter and his experts decided on the wider, more expensive tape – one of many decisions on which he would later be criticised for extravagance.

Peter likes to take a long break in the summer, chiefly to indulge his passion for sailing. In 1981 his holiday came in the middle of the equipment negotiations. When his deputy, Dick Marsh, came to take the reins for a while he decided that some of the decisions on equipment were wrong. He was especially critical of the freelance equipment consultant Peter had hired. Dick arranged a meeting with Jonathan Aitken where he expressed worry at the way Peter was running things. Then he began talks with a Japanese firm offering a package deal to equip the entire studio. That did not work out; but the consultant did leave soon afterwards and Dick's friends gave him credit for sorting out what had threatened to become a tangled mess. 'The Marsh Protectorate', they called it. Peter was less impressed, and was sore at having to clear up the remnants of the Japanese initiative. It was another flicker of a quarrel amongst the gang of friends.

Boardroom rows were rare, mainly because of the board's unusual structure. Peter, the only full-time executive on it, jealously maintained his right to be the sole spokesman for the executives. One of his early hirings had been the director of finance and company secretary, Tony Wakeling – a man with an almost permanently startled demeanour, formerly the finance director of Michael Rosenberg's medical group. In his role of company secretary he was required to attend board meetings and be available to answer questions but he was not supposed to make unsolicited contributions. Towards the end of 1981, the board were discussing how to reallocate the shares of the departing Nick Elliott. Tony volunteered a view different from that of Peter, and David Frost supported him. Peter was angry. After the meeting he reminded Tony forcefully that he must not speak unless spoken to. He explained that it was important for the board to be presented with a coherent and consistent story line from the executives and that he would deliver it. He did not want to burden them with internal differences or run-of-the-mill problems. Tony, never at ease with Peter, murmured an apology.

The notion of a single company position to put before the board was something Peter had brought with him from his Civil Service and ambassadorial days. In government departments the procedure was for the staff to discuss policy among themselves and present an agreed line – or at least an agreed set of options – to the minister. Ambassadors did the same thing for the foreign office. A related philosophy inspired Peter's instructions to his staff about writing minutes of meetings. He hated imprecise statements such as 'It was proposed that . . .' and 'It was decided that . . .' He insisted that the minutes state specifically who proposed and who decided, to encourage personal responsibility.

Another of his attempts at bureaucratic innovation was less successful. He had been keen to institute 'open' filing, again on a Whitehall model. It meant that instead of each executive keeping his or her own files, there was a common company filing system categorised by subject. In this way everyone would be able to find out what everyone else was doing. He saw common filing as the administrative equivalent of his precious windows in the studios, a device to encourage accountability and open government. But when the time came to implement it, the newly-recruited clerical staff complained that they could not understand it. They were more comfortable with their own private, locked filing cabinets. And so, it had to be confessed, were their wary executive bosses, who did not at all relish the glare of communal scrutiny. Peter withdrew gracefully.

An early secretarial slip had embarrassing consequences. Contracts for the five presenters were negotiated in 1981, and that autumn Peter sent out two copies to each of them. They were produced on a word processor. Details of all the contracts were the same except the names and salaries. Properly programmed, a word processor can take such variations in its stride, ensuring that the right salaries are allotted to the right names. It was properly programmed and produced ten immaculate and correct documents. The secretary who was to post them, however, was innocent of such advanced technology and had been imperfectly briefed. Seeing ten identically laid-out documents in the word processor's tray, she assumed they were all the same, so stuffed them into the five envelopes at random. Until then Peter had been careful not to let any presenter know how much the others were paid, fearing ultimately expensive jealousies. Just how sound that reasoning was became clear when the presenters received in the post the details of the rewards of some of their colleagues. The experienced Robert Kee, in particular, was surprised to learn that he would be earning less than Anna Ford. No immediate adjustments were made but in a review of presenters' salaries before they went on air, Robert achieved parity with Anna at £75,000 a year. They were still a long way behind David Frost and Michael Parkinson, level-pegging at £112,500 for their six-month hard grind. Poor Angela Rippon had to make do with a paltry £60,000. These figures were much speculated upon in the Press, though the guesses were not always accurate.

Towards the end of 1981 the company moved to temporary offices right opposite their new headquarters, where they could watch their future home take shape. The cost of the new building was nearly £5 million − twice what it would have cost to adapt Ewart's − and on top of that there was the equipment to be paid for. Still, Peter was reluctant to go to the shareholders and ask for more money before the station went on air. One comforting fact was that Derek Stevenson, the sales director recruited from Thames Television, was estimating first year revenues at £16 million, roughly equal to what was now being budgeted for first-year expenditure. Peter saw it as his prime job to stick to that budget, which had already gone up from £12 million. He was sure there would be pressure to increase the budget further as the start-up date neared but he was determined to resist it. At one executives' meeting he insisted that unless costs were controlled 'we'll be Murdoched' − in other words they would be vulnerable to the kind of rescue Rupert Murdoch effected on London Weekend in 1970, resulting in mass dismissals of executives and others. 'You'll regard me as a paragon of civilisation compared with what will follow,' he predicted sombrely. (A garbled version of that warning reached Dick Marsh, who rang Peter in alarm a few days later, asking about a rumour that Murdoch was bidding for TV-am.)

But budgets do have a habit of creeping upwards, despite the most rigorous policing. There are always those overlooked additional items without which the machine cannot function, those accumulated ha'porths of tar whose absence could not only spoil the ship but sink it. It was April 1982 when Tony Wakeling went to see Peter and Michael Rosenberg to say that the proposed £16½ million was inadequate. As the time approached to recruit the bulk of the staff, it was clear that insufficient provision had been made for staffing on the programme and technical side. Michael Deakin and Geoff Smith, the director of operations (and another fugitive from Yorkshire Television) kept telling him so, but Peter would not listen to them or to Tony. 'I don't want to hear you say that outside this room,' Peter snapped to his director of finance. 'I'm not going to the board for more money.'

Peter had a strict attitude to budgets. He emphasised to Tony – who found his schoolmasterly manner hard to take – that budgets were not predictions of expenditure but controls on it. He had a strict hierarchical view on this as on other matters. The board decided how much money could be spent and that was the basis of the budgets handed down to heads of department, who made their spending decisions within those budgets. If you regarded budgets as predictions of what could or should be spent, that would undermine their authority. There would be no barrier against constant upward revision. It was not that he foresaw much difficulty in raising extra money, but he was conscious that to do so would inevitably water down the stock and reduce the proportion of it held by the founders and presenters. At that time their share was around 20 per cent. If it were to go down to single figures it would be a mockery of the United Artists concept that had so inspired them all. That was why Peter insisted that Tony drop no hint to any department head that there was the faintest possibility of the budget's being increased. If he did, they would regard him as an ally and redouble their pressure on Peter to loosen his purse strings.

Jonathan Aitken was away in America at the time of the annual general meeting in July, so his irascible cousin Timothy went instead. It was preceded by a lunch at Barclay's Bank in the City. Timothy found himself sitting next to Peter and asked him casually how much money he expected to make from his shares. Peter did not answer the question but launched into an exposition of his mission to explain. It put Timothy in a sour and sceptical mood for the meeting that followed.

It was the first board meeting he had attended since early in 1981 and he was quite shocked. He heard no clear financial reporting or any evidence of rigid commercial control. The other directors had, over the months, lulled themselves into accepting standards of reporting lower than in other companies with which they were connected. They told themselves that TV-am was, after all, operating in unknown territory, its market different from that of any existing television company. Not Timothy. He sensed things going out of control when he heard what Derek Stevenson and Tony Wakeling had to say. Asked whether they would be able to keep within their budgets they replied yes . . . probably. What sort of answer was that? Hardly one to inspire confidence. It seemed to Timothy that the pair were in truth worried to distraction about the budgets and their ability to keep within them. After the meeting he buttonholed them and they confirmed his fear.

Timothy's renewed interest in the company coincided with the start of Peter's annual holiday. At the end of July he went off to Sardinia for a month – an absence he was to come to regret even more than the previous year's. Most of his senior colleagues felt that, with the station due to be launched in February, it would be risky to go away. So they stayed in London and, like the proverbial mice in the absence of their natural predator, they lost their inhibitions. Peter had such an abrupt, forbidding way with people who disagreed with him, his memoranda could be so cutting, that after a while they were afraid to express their opinions. Now he was away they could do so.

There were several separate meetings between executives and directors that August. They began with a dinner at which Derek Stevenson, Tony Wakeling, Geoff Smith and Michael Deakin all agreed they were so seriously underfunded that there might be a case for going to board members above Peter's head, risking his wrath. Michael went to see Jonathan Aitken, in hospital recovering from the broken ribs he had suffered earlier that year. Tony spoke on the phone to Jonathan and to Lord Camoys, chairman of Barclay's Merchant Bank, the main providers of capital. Geoff told Dick Marsh he was worried whether he had enough money for the equipment he needed. Jonathan, just after he left hospital, invited Derek for a drink at Lord North Street, questioning him about his revenue forecast: he feared that David Frost and Michael Rosenberg might be encouraging him to be over-optimistic. He also slipped in some questions about Peter. Was he a good manager? Was he popular? Did he seem to have a grip on things? Although Derek answered yes to all three, Jonathan, who prides himself on his psychological insights, thought he sensed unease.

Peter was back in time for the board meeting on 26 August, fit and tanned but quite unprepared for the onslaught awaiting him. Both Jonathan and Timothy were there to represent the Aitken interest, but Timothy did all the speaking. He had been well briefed on all the August discussions and now, unlike in June, he had something concrete to base his complaints on. It was plain to him that there was not enough money in the budget and, most seriously, no cushion of reserve financing to safeguard against early setbacks. He detected nothing except a blithe optimism that it would be all right on the night, and suggested that management energy was being expended in the wrong areas. Peter was paying too little attention to financial control and too much to the creation of programmes (an observation that echoed Michael Deakin's repeated complaints about editorial interference). Peter insisted that the reverse was true and quoted one of his textbook management philosophies. 'Finance is after all the unifying element central to all management disciplines.'

Timothy recommended two courses of action. The first, viewed with sympathy by many board members and Peter himself, was the appointment of a finance director of greater experience than Tony Wakeling. The second suggestion shocked some of the directors – notably the conservative city figures Jacob Rothschild and Lord Camoys. Why not, Timothy wondered, offer shares of TV-am for public sale in the Unlisted Securities Market? Jonathan had mentioned the possibility at an earlier meeting but he had done it so diffidently that the others thought he might be making a joke. Timothy certainly was not joking. Rather, he seemed to become more and more seized by the idea as he spoke. He believes in raising capital on the open market because it subjects companies to the most rigorous discipline and scrutiny. By introducing the element of accountability, it allows no scope for laxity in financial control. Proper information has to be given.

The other directors were horrified at what they saw as the irresponsibility of the suggestion. How could you ask the public to invest in so highly speculative a venture, with an untested market and (although they did not put it so bluntly) untried people at the helm? They thought it so outrageous that they wondered if there was anything behind it. Did the Aitkens want a capital value put on their shares so that they could pull out if necessary? Or did they want to buy more shares and move towards a greater degree of control? They had, after all, originally sought a 20 per cent holding rather than the 15 per cent they were allotted. In any case, on purely practical grounds, the hard-pressed TV-am staff could hardly be expected to do the enormous amount of preparatory work a public listing would require.

Many of the other directors were resentful at the way Timothy – no more than an alternate director for his cousin, after all – was throwing his weight about. Until then it had been rather a cosy board. They would sit and listen, almost mesmerised, to Peter's long and reasoned accounts of the latest doings, the people he had hired and was planning to hire, the progress of the building. It was all civilised and companionable. A few questions would be asked but they would be constructive, with no hint of rancour. And now here was this aggressive young man speaking out in the most blunt terms, even implying that Peter was not competent for the job. The impudence!

Peter, however, took it all in his familiar stoic fashion. Not for him to bang on the table and cry 'Nonsense!' If he seemed taken aback when Timothy began, it was only momentary and he soon regained possession of himself. He said he had not realised the extent of the doubts about the budget figures and it was certainly something he would have to look into. He was puzzled why this had not been raised with him in detail before. He thought he ran an open organisation, where people were encouraged to speak their minds. He would consult his senior executives as a matter of urgency and go through the figures again in depth, then report back to the board at the following meeting.

It had been an unhappy morning. Two years earlier, when the gang and the money men were all so keen and supportive of each other as they prepared for the IBA hearing, such a bitter argument would have beeen unthinkable. They were such good friends. Now, as the meeting broke up, eight board members went off for lunch at Le Routier in Camden Lock, then the nearest decent restaurant to their headquarters. But it was no celebration. They did not go in a group, as they had done on that triumphant December day at the Hyde Park Hotel. Now the eight went as four distinct pairs, sitting at separate tables. Peter and Jennie sat outdoors. The other six went inside, to tables strategically distanced from each other. It was a time for quiet, earnest and private talking.

Next day, 27 August, Peter sent a memorandum to his senior colleagues, saying that the figures agreed in the early summer now had to be reviewed. An intensive series of discussions followed and some horrifying gaps in the plans became apparent. Geoff Smith, when he joined as director of operations in 1981, had been told he would have a budget of £6 million for the initial installation of equipment. Peter, wielding his knife, had made him settle for £4½ million, but it was now clear to Smith and to Geoff Monks, his chief engineer, that without at least an extra £½ million the station would never go on air at all. Smith also insisted that in order to avoid exhausting and expensive overtime he would need to increase his staff complement from 169 to 256.

It is a natural instinct for apprehensive executives, about to launch into some new and unpredictable venture, to want to pad themselves with extra staff, but Peter's inquiries uncovered more specific omissions. For instance, no provision had been made in the budgets for the maintenance of an editorial library, an essential service for any news-gathering operation, and labour-intensive because it has to be kept up to date with newspaper clippings every day. Geoff thought this fell inside Michael Deakin's programme department but Michael had assumed it was covered in Geoff's budget.

Two weeks later, Peter sat down and compiled his sackcloth-and-ashes statement for the board. It was headed: 'Budget

Review – Interim Report by the Chief Executive' and dated 17 September. Specifically, it recommended that the budget should now be increased by more than £3 million to £19.7 million. Furthermore, after consultation with Derek Stevenson, he had to revise his revenue forecast. It would now take longer than anticipated to build the company to profitability but that happy day still ought to occur some time in 1985. He recommended a two-year postponement of the repayment of loan stock, until 1988-9.

Peter now believed that the realistic cost of what had been proposed in the budget calculated in May was £22.9 million, but he was not going to ask for an increase as great as that. Some £3 million could be saved by trimming those proposals but to get them back to £16½ million would result in a cut-price service that violated many of the commitments given to the IBA. Confessing that this was a grave position to report to any board, Peter said it provoked six questions, of which the three main ones were: Why were the May figures wrong? Will the September figures be more reliable? Is management adequate to its task? Lumping the first and third questions together, he made his frank confession of failure at the highest level, and took full responsibility for it. Elaborating, he explained that he had not recruited enough senior management to give substance and realism to the May budget, that he had not ensured that everyone fully understood everyone else's plans and that he had not been sufficiently sympathetic to pleas for higher expenditure from heads of department.

Peter absolved Tony Wakeling from responsibility for the debacle. There had been no lack of tight financial control: indeed the very tightness of it probably contributed to the problem. 'I plan to strengthen management and support systems in a number of ways,' he pledged.

Such an unspecific undertaking did not, however, satisfy Timothy, who was beginning to believe that the appointment of a strong finance director was not the limit of the changes needed in senior management. It was not the figures they should be looking at but the people, he stormed. Were they competent to run the company? The board were given figures that only three months later proved to be inaccurate. As for Peter's paper, you needed a degree in economics to understand it. Perhaps the old plan for a managing director to share power with Peter needed to be revived, but as a first step Timothy proposed a thorough audit of the figures by Arthur Andersen, a respected firm of city auditors.

Peter opposed this. He stated that auditors, concerned with making the books balance rather than with where any extra capital would come from, were unlikely to strengthen the case for economy. He made this point as rationally as ever, but after his confession of error his standing among many of the institutional directors was diminished. They were no longer disarmed by his eloquence. For almost the first time, they were seriously worried about the running of the company. They had known all along that this was a risk investment but had seen the risk as inherent in the creation of a new business. Now it was being compounded by admitted incompetence at the top. If Peter believed that his *mea culpa* statement would end the matter, that the directors would pat him on the head, telling him to forget all about it and try harder next time, he reckoned without the crucial factor that it was their institutions' money he was risking. Nothing discourages sentiment, or crushes the schoolboy notion of fair play, more than the vision of millions of pounds floating through a boardroom window. The board approved Timothy's suggestion of commissioning an auditors' report.

The report, presented at the October board meeting, was written by Ian Hay Davidson, now chairman of Lloyd's underwriters. He came to broadly the same conclusion on the level of the budget as Peter and Tony Wakeling had in their revised figures in September. More pointedly, Davidson implied serious weaknesses in the company's management structure. His report strengthened support among the directors for Timothy's view that Peter had too much power and should share his authority with a chief executive.

This was a heavy personal blow for Peter. He had invested nearly two years of his life in running this company. Although the board had never formally rescinded their plan to have a separate managing director eventually, in his own mind Peter had come to think of himself as the sole head until well beyond the start of transmission. After all, he had no other job. He was not at all sure that, having held supreme power for so long, he was ready to relinquish an important part of it. He put up an alternative proposal to the board. Why not appoint a tough and experienced finance director, as Timothy had suggested earlier? That would give him the support he needed in the area where he was perceived as being weakest, without muddling the lines of authority by importing someone to share power.

Peter was sent out of the room while the directors discussed what to do. He walked from the temporary Camden Town building, where the meeting was held, across Hawley Crescent into the studio building. Two weeks earlier he and the programme department had moved their offices there, although the place was still packed with builders, decorators and engineers rushing to finish it. He was pleased with the building – functional and fresh. How ironic if he were not after all there to see the first programme broadcast from it. For if the board, in their secret conclave, decided to make him merely a figurehead chairman, he did not really see how he could stay on.

Peter had some supporters at the meeting but none spoke so forcefully and persuasively as his opponents, the Aitkens, urging the board to replace him as chief executive. The institutional investors were on their side, believing in the efficacy of experienced management. They had an ally in Michael Deakin, who made a surprisingly fiery speech criticising Peter. He mentioned the disputes about editorial prerogative and complained that Peter had written a letter direct to members of Michael's staff instead of going through him. Some directors felt this was an unfair attack on a man who, not being in the room, could not answer back. The board decided to accept the recommendation about a chief executive. Jonathan Aitken was put in charge of a sub-committee to find one. Dick Marsh, a vociferous supporter of the move, was deputed to walk across the road and tell Peter. Peter said he wanted to think about it over the weekend.

It was a weekend of many telephone calls and meetings as Peter tried to limit the damage that was about to be done to his position. His chief ally was David Frost, who feared that the whole enterprise would be ruined if Peter, feeling humiliated, decided to pull out completely. Together they went to see Jonathan at his house in Lord North Street. Their objective was to agree terms of reference for the new appointment that would make it clear that the person to be appointed would be Peter's subordinate, with the title not of chief executive but managing director. Peter, as executive chairman, would thus still be indisputably in charge. Jonathan argued, in his measured manner, that this was not, as he saw it, what the board had agreed. They had specifically wanted someone to run things because they thought Peter had been deficient in doing so. Nobody could run a company as a number two. He envisaged a structure where the new person would be the captain of the ship and Peter the admiral, removed from day-to-day administration.

No agreement was reached at that meeting and the negotiations continued for a few days. Meanwhile David had, in the greatest secrecy, been to see Lord Thomson at the IBA to discuss the moves against Peter. Thomson told him that Peter had been an integral part of the group to which the franchise had originally been awarded and the IBA would be concerned if he were to pull out entirely. But he was clearly in no position to give any formal declaration of the IBA's support and did not do so.

In the end Jonathan put on paper the relationship as he saw it between Peter and the proposed new managing director. As executive chairman, Peter would be the head of the company as seen by the world at large, dealing with external relations, particularly relations with the IBA. The managing director would run the internal administration, including finance. Any dispute over their respective powers would be resolved by the board.

It was a victory for Peter and David. What the board had originally decided had been watered down. The purported chief executive would now be a managing director. A firm of headhunters was appointed to search for the right person. Peter addressed his staff – the nucleus that had so far been appointed – and told them of the change. In doing so, he not only put into practice his notion of open government, but helped ensure that those hard-won terms of reference were not altered again.

You win some, you lose some. While Peter triumphed in the matter of his status he lost the battle of the budget. In his September paper, Peter had recommended a budget of £19.5 million, higher than the May figure but still over £3 million below what he calculated it would cost to implement all the plans agreed in May. The Aitkens wanted the full £22.9 million budgeted and the auditors agreed with them, as Peter had predicted they would. To support his argument Timothy had brought with him to the board meeting a survey from a city analyst on the prospects for breakfast television. This estimated annual revenues of £25 million, higher than the previous most optimistic forecasts. He took it in his fist and brandished it at his fellow directors as proof that a budget of £22.5 million was perfectly sustainable.

Other reports had drawn notably different conclusions. Jennie Bland had a copy of an article by Harry Henry, a marketing specialist, in the magazine *Admap* the previous October. Called 'Breakfast Television: Much Ado About Remarkably Little', it took a highly pessimistic view of the prospects. Based on American and other experience, Henry estimated that the total audience for morning television would be no more than 3 per cent of the population. If the BBC were to compete, that would leave TV-am with just 1½ per cent. In revenue terms, Henry calculated, the best they could expect would be £9.2 million.

While Jennie was explaining this to the board, Timothy interrupted her to ask how many shares she owned. She confessed to her paltry holding and Timothy stormed: 'I should like to put the point of view of one of the major shareholders.' He explained that Aitken Hume would not like to be associated with anything that did not 'go with a bang'. (The explosion that eventually occurred might not have been what he had in mind.) The show, he said, had to be punchy, and if that cost money then it had to be spent. Above all, the programmes must be a success. This view gained sympathy round the table. Jennie pointed out cautiously that it was all very well hiring extra people at great expense, but it was often the devil's own job to get rid of them. Timothy said that risk had to be taken. He insisted that the people who had produced the forecasts for him knew what they were talking about. Swayed by his passionate advocacy, the board agreed that the budget should be £22½ million. Peter ensured that his objections were recorded in the minutes and declared that despite the vote he was going to try to restrict spending to the £19½ million he had been advocating.

Some board members, even those who supported Peter, felt that the seeds of the budget crisis had been sown more than a year earlier, when the attitude to spending had been a lot less disciplined. Terry Farrell, the architect, thought that the cost of his conversion of the Camden Town garage – just above £5 million – was not at all expensive, but some directors thought it could have been done cheaper. They detected that attitude of carelessness about small – and by themselves insignificant – items of expenditure that is a hallmark of lax financial control. Peter Jay's office, for instance, was going to cost £7,000 to decorate in the original plan, with the offices of lesser executives only a little cheaper. The estimate for interior decorating seemed high at £800,000.

The most contentious item of all was the small house on two floors squeezed into a corner of the building facing the canal. Terry Farrell was keen to put it in and Peter thought it would make an ideal *pied-à-terre* for him on those occasions he had to stay late, as well as a place to accommodate important visitors who were appearing on the programme. There were two small bedrooms, a kitchen and, on the first floor, a tall living room, with high picture windows, that would have provided a stimulating venue for informal meetings of executives. The projected cost was £40,000.

When the building plans first came before the board they vetoed the house on the grounds of cost and were surprised to learn some months later that it was nevertheless being built. They were told it was not going to be paid for out of the TV-am budget but Michael Rosenberg's company would build it and then lease it back to TV-am. Michael was a wizard with lease-back arrangements but this one fell through and the house went back into the building budget, although the directors were not informed. And the cost doubled because of structural difficulties with the foundations.

Jennie Bland has an eye for such housekeeping items and raised them at board meetings. When she did, other directors would show impatience that time was being wasted on details.

Some such expenditure never went before the board at all. The purchase of a boat on the canal as a place of relaxation for staff came as a surprise to the directors when it was disclosed in the press. So did the giant egg-cups on the roof. These cost very little – £1,200 altogether – and made the building a talking point, but some thought they looked vulgar and did nothing for the local landscape. The technical equipment cost £5 million. This was approved by the board without question simply because no director had any knowledge that would enable him to query what was needed or its cost. More gravely, Peter did not have that knowledge either.

The building has a metallic grey, horizontally-layered fascia, curving with the street, on either side of a high-tech neon-lit entrance arch banded with orange and yellow piping. Inside, the reception area leads into the atrium which Farrell designed to suggest a Mediterranean garden within a desert, complete with indoor yucca trees costing £16,000. The staircase he envisaged as a Mesopotamian island and the hospitality suite a Japanese tea pavilion. The architect also designed the furniture. The multi-coloured chairs with fake wheels hark back to the traditional seats of Hollywood film directors.

All that makes it sound more elaborate than it looks. Unless told about the desert and the tea pavilion, any but the most imaginative and style-conscious visitor sees it as a fairly routine conversion of an old garage, with a few trees and plants and some ultra-modern decor. Farrell insists that despite the gimmicks the building is an object lesson in economy.

The trouble is not that it looks extravagant as a whole, but that the quirky pieces of individual styling give a fake impression of frivolity, of a carelessness about expense. When – as was to happen with TV-am – the executives are criticised for being wasteful, it is hard to sustain the argument that a building with egg-cups on the roof, yuccas, glass walls and a Mesopotamian staircase inside is no more than functional. Managers who want to impress shareholders with their parsimony are better advised to work from austere, charmless square boxes without extraneous frippery.

With the building nearing completion and the equipment ordered, all that remained was to fill it with people. A staff of

430 was needed to perform the programming, technical, administrative and clerical functions that would get the station on the air. How to recruit them? Peter turned for advice to two organisations – the London Business School and Saatchi & Saatchi, the advertising agency that had burnished the Conservative Party's image in the triumphant 1979 election campaign, now appointed by Peter to create TV-am's advertising. The business school drew up a long recruiting questionnaire based on the most up-to-date scientific research, while Saatchi & Saatchi placed a full-page job advertisement in the *Guardian*.

The result was 13,000 letters of application, mainly from people without relevant experience. Some had just left college, or even school, while others had worked all their life as brush salesmen and their wives had told them they had just the face for television. There was only one full-time personnel administrator, helped by two temporary clerks. They had to send out application forms, then read and process them when they came back.

The questionnaire, eight pages of it, did not help. The questions were detailed to the point of being intrusive. Some would have been appropriate on a psychiatrist's couch, others in a survey on working conditions and industrial relations:

How does a move to TV-am fit in with what you have been doing?
What particular talents or interests, other than your formal qualifications, would you bring to this job?
To what extent are your formal qualifications used in your present job?
To what extent are your talents and interests used in your present job and how would you see these being harnessed by TV-am?

Applicants were asked to give in detail their hours of work on each day of their last full working week. Was this normal or exceptional? If exceptional, why? How much time was spent waiting for instructions, briefing or clarification?

Quite a lot of the personnel manager's time was, in the event, spent on clarification, for some respondents found the questions so difficult that they had to telephone to find out how they should answer. On average, three people were interviewed for every vacancy and it is scarcely polite to see them for less than half an hour each. For weeks, Michael Deakin and Geoff Smith had time for scarcely anything else. And yet, Michael reflected ruefully, there was Peter still fussing about getting a blueprint for the programmes down on paper. That man, honestly! He seemed to think TV-am was a branch of the Civil Service. Still, on the credit side, Michael was managing to recruit, he thought, some exciting people. Not hard-bitten, cynical professionals but the best and brightest young women and men from the universities, with supple minds and dazzlingly sharp responses. He looked forward to creating programmes with them as keenly as a young suitor looks forward to making babies with his clean-limbed virgin bride. Maybe he would have them in his office for rap sessions. They would sit around with coffee in paper cups. Ideas would fizz through the air, as they had done when he worked for Yorkshire. That, thought Michael, is how first-rate television is made.

CHAPTER 4

The Talented And
The Nice

'Good afternoon. I'm Peter Jay; and I'm here to tell you in as few words as I am capable of – which my dear friends and fans say is not a million miles from millions – what essentially TV-am are all about.'

During November the newly-recruited staff arrived in weekly waves. For those on the programme side, an induction routine was devised that began with an address by Peter in the temporary offices across the road, now nearly empty because all departments had moved into the new studio building. The recruits were young, keen, extraordinarily excited and prepared to receive Peter's dressing-room pep talk in the gung ho spirit in which it was delivered, before rushing on to the field for the kickoff.

But first, let me say a very warm welcome to everyone. The next weeks and months are going to be among the most hectic and demanding of your lives. We are asking one hell of a lot of you. In return and above all I want you to have fun, to enjoy it, to have lots of adrenalin released by the challenges of doing something totally new, sometimes frustrating, sometimes impossible, but always in the end fulfilling because it is exciting, because it is worthwhile, because you each and every one know that you personally have a key and real role to play and because you are doing it with friends, all of whom are as nice and as talented as you are yourselves. . . .

So what is TV-am about? Essentially, I think we are about five things:-

111

1 About making television at a new time of day;
2 About making a new and different kind of television;
3 About making a new and different kind of journalism;
4 About putting into practice a new and different kind of management style and philosophy; and
5 About being a new and different kind of ITV company. . . .

Our overriding editorial task is to ensure that by their tone, pace, style, content and timing our programmes infiltrate themselves into the often very rigid start-the-day rituals of our audience so that "Good Morning Britain" [the title of the main programme] becomes as indispensable as a first cup of coffee, a first glance at the papers or – perish the thought – that first cigarette. But for that very reason we are, in *this* respect, building on the very long-established habits of the legendarily conservative British public, namely their addiction for over a century to the tradition of the popular newspaper – that special blend of news, useful information and fun – as the proper accompaniment to breakfast and the start of the day. . . .

My third point – a new and different kind of journalism – brings me back to what was, I now think, the single most important idea that won the breakfast franchise for TV-am, namely our "mission to explain". We have of course an equal mission to entertain – not by traditional "light entertainment" but by entertainingly informing and diverting our audience.

The mission to explain, to be sure, does no more than emphasise the third of the three co-equal functions of all good reporting, namely to find out, to report and to explain. And I will not now add, you may be relieved to hear, to the billions of words that I already seem to have spilled over the last three years and more about this notion. You are all welcome, if you want it, to copies of my now notorious lecture on the subject, "What is News?", the tedium of which Michael Deakin claims has stunned more sheep at a greater range than even the works of Hegel and Channel Four combined. I will only say that in everything we do the aim should be not to impress the audience – still less each other or other journalists – with our knowledge and sophistication, but to reach out for that wonderful moment when the light of understanding is lit in the eye of the viewer as he or she, old or young, says: "Oh, *now* I see". . . .

'And the new and different management style? These seven words are mine, not Jay's and should be differentiated. We can and should, I believe, together achieve – by openness, by mutual trust, by sharing of experiences and ideas and by an atmosphere of personal equality – a real sense that it is our company, all of ours, that everyone counts in it and that its achievements and its problems belong to us all. I realise of course that first and foremost we who are already here, the founders and the management, have to prove to you in action that we really mean all that . . . to help close the all-too-easy gap between lofty good intentions and practical realities at the coal-face. . . .

'Our viewing bridge – the one personal whim that I have imposed on the architect – enables everyone (though not, I suppose, at the same time) to see and to feel part of what is going on in the control rooms and the studios without interfering with their operation. And that includes guests and visitors too.

'One plea I make to you. However enthusiastic and tolerant you are – as I know you are, or you would not be here – about this new frontier of television journalism, there are initially going to be frustrations, flaws in the organisation, moments of seeming chaos, the pressure to panic. Please try to see these in proportion, to forgive the apparent culprits, to refrain from taking it out on each other and to recognise that the other guy is probably trying his or her best too.

'Remember above all that together we can – being so nice and so talented – work it out, that if the crew start bashing each other over the heads with their paddles, the canoes will not get down the rapids intact. So when the falls get steep please don't try to swim for it. Just keep cool and keep paddling. . . .

'And now I will detain you no longer. From now on we need action, not speeches. But remember it is supposed to be fun and, if it is fun, the hard work will seem easier and will be more fruitful. Think of TV-am in the first person singular and the first person plural – not us and them but me and us. It is you, we together, who are TV-am and who are going to make breakfast television a smashing success, proving yet again to the cynics and faint-hearts that there is nothing which talented people, given the opportunity and the trust and backing of their leaders, cannot achieve.

'So go to it, work like hell *and* enjoy it. One day when you are old like me you will look back with pride and pleasure on the early days, when we built TV-am from nothing; and you will say, I predict: "It's odd, you know, I've never worked so hard in my life; and I've never had such a good time" '.

For some, the good times stopped at about that point, for they had to draw right away on that part of Peter's speech that exhorted them to keep a sense of proportion about the frustrations and chaos. It is unlikely that he was referring specifically to the induction routine that the conscripts now had to endure, but they felt he might have been. Nobody seemed to know what they were supposed to be doing or where they ought to be doing it. Scheduled instruction sessions did not take place. To mix Peter's metaphors, many of those who had taken their paddles to the coal face felt provoked, from the very beginning, to seek a suitable head to batter with them.

In long discussions before the bulk of the new staff arrived, a formula for the weekday programmes had been agreed. The first hour, between 6 and 7 a.m., would be called 'Daybreak'. This would be devoted largely to news with bulletins repeated every few minutes, although using different snatches of film if possible to give some variety. This was intended as a factual service for early risers and would include weather, traffic and public transport information. The main showcase programme, 'Good Morning Britain', was to follow from 7 to 9.15. Hosted by two of the famous five presenters, in monthly changing pairs, this would contain some regular features, filmed in advance, but was essentially envisaged as a news (or 'news-reactive' in the house jargon) magazine. Peter was still using his comparison with a morning-after-the-election programme, with news constantly flowing in. Now he added a fresh example, the landing of a space shuttle, with commentary from the spot backed up by studio interviews with experts.

One of the trickiest details to be settled, requiring all Peter's powers of diplomatic negotiation, was the duty roster for the presenters. It was complicated by their various alternative commitments during the six months when, by their contracts, they were not required to work for TV-am. Michael Parkinson, for example, had a full six-month summer stint arranged in Australia, so he wanted to be working at TV-am in a single stretch. David Frost had several assignments in America to fit in. Some wanted one month on and one month off, others three months on and three off. It was a nightmare. When he did finally slot it all into place, Peter was so relieved that he had it promulgated (his word) under his signature, not to vest in it the authority of immutable Holy Writ but so that if there were to be later changes they, too, could only be done on his personal authority, by means of a fresh promulgation. Michael Deakin was delighted at the thought of

Peter promulgating a roster – an expression he felt must derive from Winchester or the Treasury. But aside from its satirical potential, the business did worry him slightly. As director of programmes, was that not *his* area of responsibility? Peter said no. The presenters were founders, shareholders and among the company's prime assets. It was only proper that they should deal on such matters directly with him. Michael fretted but did not make a scene.

The rosters for the less exalted personnel were uncomplicated by outside commitments and were decided at a lower level. But they were still the subject of a dispute that went to the heart of the indecision about exactly what kind of programmes TV-am would put out. Reacting to the news is a labour-intensive business. It needs a full complement of reactive reporters to be on duty through the night, and a camera crew ready to broadcast their reaction. And although Geoff Smith, the director of operations, had managed to abolish the most excessive 'golden time' payments for technical staff, they still had to be paid a premium rate for such unsocial hours.

Resources are finite. If you deploy your reporters and camera crews at night it reduces the number available during the day. That is how a sharp rivalry began to develop between the news people and those who made the feature material. It was vital to have a reserve of timeless items to slot in when the news flagged, but the features side complained that they were being deprived of essential resources by the greed of the news people.

On one side were ranged Hilary Lawson and Kevin Sim, the features editor, another recruit from Yorkshire TV. Michael Deakin, by instinct and training a features man, supported them, completing the 'Yorkshire Mafia'. In fact that was a misleading designation because one of the two cheerleaders for news – managing editor Clive Jones – was also a Yorkshire alumnus. He and the news editor, Bob Hunter, were constantly agitating for more resources to be devoted to the news side.

Hunter, from Northern Ireland, had been recruited from ITN on the recommendation of Robert Kee and Anna Ford, who had both worked with him. But from the start he did not get on at all well with Michael Deakin. Michael delights in ascribing such disagreements to the inherent ethnic and psychological characteristics of those who oppose him. He quickly cast Hunter as a morose depressive, affected by Celtic gloom, an incorrigible conspirator, and he would so inform anyone who cared to listen. Hunter's position had been additionally complicated by an early disagreement over his working relationship with Hilary Lawson, the editor-in-chief. Hilary was notionally Hunter's superior but in order to assert the quasi-independent status of the news side Hunter insisted that he have direct access to Michael, skirting Hilary. On the chart Michael circulated to illustrate these hierarchical niceties, Hunter appeared below Hilary but a broken line led directly from Hunter to Michael, indicating their special relationship. Such ambiguities were characteristic of Michael's administrative technique. He liked, if he could, to accommodate everyone's susceptibilities, even if they were patently incompatible. It was an infallible recipe for misunderstanding. Despite the broken line, Hunter was never part of Michael's inner circle, the Yorkshire Mafia. When he decided to quit in December (though he stayed until mid-February), it was put about that he had gone because he thought news was getting short shrift in the programmes that were being developed. That was indubitably part of the reason but a more important cause was his personality clash with Michael. It was a serious loss at a critical stage in the company's development.

With egos already colliding so sonorously, it was difficult for the presenters to make their voices heard on the question of programme content. A programme committee was established, including the presenters and senior programme editors. This was to be the channel for that mysterious 'input' that the presenters were supposed to enjoy on questions of content but it seldom met because of the incompatible shift patterns worked by its members. In the booklet given to joining staff to explain how the company operated, it was stressed that the programme committee 'does not make decisions'.

Just who did make the decisions was among the deep mysteries the new recruits had to unravel. They had expected Hilary to be the fount of ideas and instructions, but he was disconcertingly reticent. In the early weeks he called numerous meetings of producers. They hoped he would use the occasions to prime them with instructions, but instead he would ask them what ideas they had thought of, and leave them to get on with putting flesh on the skeleton. Chatting amongst themselves, the producers would wonder whether Hilary was out of his depth. Michael Deakin, sensing their unhappiness, called them to his office for pep talks. Once the journalists took the initiative and arranged a meeting with Peter, Michael and Hilary to express their unease. 'Don't worry,' was the message they received. 'It will work out all right in the end.' But would it?

The sharing of experiences and ideas, which Peter had made so much of in his opening pep talk, thus proved elusive in practice. Instead, the staff felt they were wedged in that uneasy gap he also mentioned, between good intentions and practical realities. The intentions were symbolised by the preponderance of glass walls in the new building, not just around the studios but also enclosing his office and those of his senior executives. Through the glass the toilers could see him working and admire his other wall, on which were displayed scores of trophies and personal mementoes. There were photographs of himself with President Carter or in his sailing boat, certificates of honorific awards, including the freedom of numerous American cities, presented to him during his term as ambassador. Later, in attempts to pour scorn on Peter, this memorial wall was adduced by his foes as evidence of self-aggrandisement, along with his other foible of taping every interview he gave and having the verbatim record typed up. In fact, as he had told Irma Kurtz in an interview for *Cosmopolitan* a year earlier, the office wall was an elaborate joke, based on his experiences in America, where no executive suite is complete without its display of certificates and signed photographs. 'Having spent a lot of time in American Congressmen's offices I always thought what fun it was when you were waiting to see somebody that you could look at these exotic pictures on the walls.'

It is characteristic of Peter that he would automatically expect everyone to share the joke even without the benefit of his Washington experience. For although he genuinely believes in the principle of democracy in office relationships, he is innocently unaware that he is scarcely the right person to put the principle into effect. He does not appreciate how forbidding a man he is, with his severe intellect and his flow of

117

rationalisation ready to be switched on the instant he is apprised of a difficulty, making his listeners feel fools for not having thought of his solution. A collectivist in theory, he is by instinct the sternest of autocrats. Tony Wakeling was not the only one of his subordinates to be overawed. Peter did not understand how unthinkable it would be for any of them to drop by for a casual chat. If they wanted to divine his thoughts, they were left with trying to grapple with his forty-eight-page 'What is News?' lecture and with its daunting conclusion:

I believe that decent news journalism can survive and prosper, true to its historic ideals – fearless, fair, truthful and illuminating – winning and holding public confidence in and support for its essential freedoms and building for itself as a profession an ethic, a self-confidence and a set of standards which can liberate it from some of the tawdriness, the amateurism, the self-satisfaction, the self-indulgence and the pretended immunity from the imperatives of humanity and citizenship which in this country have too often provided ammunition for the enemies of freedom and which have shocked and alienated the very public whom it is our duty to serve and in whose name alone we are entitled to claim the necessary privileges and immunities of our profession.

Hard-hitting stuff, but not likely to endear him to those members of his staff who sprang from the tradition he was describing as tawdry, amateur and self-satisfied. Chief among them was Bob Hunter, who produced what was seen by some of his staff as a counterblast to 'What is News?', a more practical guide for reporters, its one sop to Peter being the recommendation that every news report should incorporate a sentence or two of explanatory material.

Despite these random attempts to articulate a TV-am philosophy, the new staff still felt bereft of guidance. Even the best and brightest required something substantive to exercise their first-class minds on. It should have come from Michael Deakin but he felt inhibited by the existence of the programme committee, which seldom met. The presenters were supposed to be helping decide what kind of programmes they should be putting out, but they had no consistent view. Was it news leavened by interviews and commentary, or was it essentially a topical talk show plus frothy features with hard news pushed into the background? Michael Parkinson and David, coming from years as hosts of talk shows, favoured the latter construction. So did Michael Deakin, with his ingrained preference for glitter over solemnity. But Anna's and Robert's experience was in hard news reporting, while Angela appeared ambivalent. If the presenters were indeed to be allowed to do it their way, Hilary Lawson and his production team were faced with the nightmare prospect of changing the show's identity every time the roster threw up a new combination of presenters. Hilary could never truly be sure whether he was working for the presenters or they for him. Conflicts were shelved, not solved. Only Peter had the ultimate power of decision-making and not every little quarrel could be referred for his arbitration. Enough were, though, to test beyond tolerance his theory that good will and an intelligent approach could heal any differences. On one point, at least, everyone was united – that Angela and Anna should not be stuck with presenting lightweight 'girlish' items. All the same, Anna was far from happy at the way things were going and towards the end of 1982 thought seriously of resigning. She wrote to Peter about it and was persuaded to stay on by an immensely long letter he wrote in reply.

With the essential question of programme philosophy unresolved, Hilary's advance schedules could only be tentative. He would describe them as bread to cast on the water to keep the ducks contented. This was a small personal joke of the kind he and Michael Deakin enjoyed. Michael kept an illuminated plastic duck in his office and was fond of jokes involving small creatures. He presented Clive Jones, the managing editor, with a hamster on a treadmill, symbolising the remorseless nature of his task. When Clive pressed him unduly on any matter he would observe: 'Clive is hamstering me to get this done.' He may have used his jokes to shield himself from assault on matters where he felt vulnerable. Another of his idiosyncrasies was to jot down notes on the developing crises.

The outcome of the rivalry between news and features was an unsatisfactory compromise. The staffing pattern established was as for a news operation, with four shifts of duty round the clock. In practice, because not much news happens at night, the duty reporters would often sit idle. Yet there was an insatiable demand for timeless 'fillers' from a features department seriously deprived of resources. There were many arguments over facilities, most vitally over the gravely overstretched editing capacity.

Editing is one of many aspects of television over which passionate doctrinal disputes rage. The argument is about whether the people who operate the machines should be technicians who know how to fix them when they go wrong but are deficient in the artistic niceties of editing; or whether they should be professional editors who would do a more polished job but do not know what to do if a bulb goes out. Customarily, the programme people like to have professional editors – but that means also having engineers on standby to fix breakdowns twenty-four hours a day. In a characteristic compromise, TV-am decided to hire a mixture of both; but to make a decision is not to implement it. Tape editors are much in demand in the industry, not easy to find, and in the early days they had not found enough.

This meant that reporters who came back into the studios with a taped item in the afternoon would sometimes have to wait the whole evening before they could get it edited. Peter's lofty philosophies were of little help to them in overcoming that singular frustration. They and the tape editors and the engineers would work hours of highly-paid overtime. There were stories that some categories of engineer were racking up overtime at the rate of £500 a week.

The editing facilities themselves came in for criticism. Editing is customarily done in individual suites, each containing a single video tape player. This is because the sound as well as the visual material has to be edited and if all the machines are in a common area a confusing cacophony is created: one sound-track becomes hard to distinguish from another. Yet this was the way it was done at TV-am, to everyone's irritation.

The motive was not perversity. There were two reasons. The first was the shortage of qualified editing personnel: having all the machines together meant that one person could look after several at once. The second sprang from the fact that TV-am was to be a nationwide service. Originally the IBA had intended that it should run only national advertising, but one of Peter's diplomatic successes in negotiating the contract had been to get them to agree to allow advertising to be sold on a regional basis, against bitter opposition from the existing independent companies, worried by the competition. This increased the pool of potential advertising. While some national advertisers want their message relayed to the entire country at the same time, many are more interested in the limited regional market. But while the concession presented Derek Stevenson, the director of sales, with a valuable commercial opportunity, it confrónted the technical people with a unique challenge. For the company had no facility for transmitting selectively to the regions. All the programmes and the commercials originated from London. So a device was needed that would send up to seven different commercials simultaneously to the seven regions into which the country had for this purpose been split. No suitable machine existed, so one had to be invented.

GEC-Marconi, the electronics firm, set to work to develop it, co-operating closely with TV-am's engineers. It was clear that it would have to be a computer of great sophistication. It was also clear that it would need to be wired to a bank of tape-running machines sited side by side in a common area. These could be the same machines used for editing. So although the open-plan editing suite made life difficult for the programme makers, it was vital for the commercial department.

The new computer, called Protel, cost £500,000 and took longer to perfect than Marconi had envisaged. The original plan had been to phase in regional advertising after some months of operating on a national basis but under pressure from Derek Stevenson the date was advanced to late in February, the start-up month. But as the start day approached, it became obvious that it was going to be a race to get Protel ready. The problem consumed more and more of Peter's time during the last months of 1982. He chivvied Marconi as much as he felt advisable but he thought too much pressure would be counter-productive. In the end, it came down to a single technician or group of technicians working to get the machine into shape. The more he nagged them, the more they would have to break off from that work to make their excuses to him.

He was convinced they were as keen to keep to their schedule as he was.

But keenness cannot by itself solve complex technical problems and by January it was clear that Protel would not be ready by opening day on 1 February. Peter's dilemma now was that the computer had been designed as the only system for relaying commercials, even the nationally transmitted ones. Without it the station could broadcast no advertising at all. So expensive alternative equipment had to be hired from a video services company for sending out the ads. In the event it was not until early summer, after the managerial crisis, that Protel was ready and regional advertising could be accepted.

With brand-new technology snags are to be expected, but there were no excuses for the administrative foul-ups. Peter could see from an early stage that Michael Deakin's talents did not run to administration, especially where finance was involved. Programme makers and news reporters need access to cash for transport and other expenses but, so far as Peter could see, no properly controlled system had been set up to provide it or to oversee other programme expenditure. On 23 December many of the staff had been to a Christmas party in a local wine bar. Tony Wakeling returned to the office after 5 o'clock and received a summons from Peter. He wanted to move Wakeling's financial accountant, Geoff Rutland, from the finance department to the programme department. 'They're totally disorganised and if we don't have him in the programme department we won't go on the air,' Peter said. Although Wakeling could scarcely afford to lose him, he agreed.

On the credit side, Derek Stevenson was encouraged by the response he was getting from advertisers, eager at least to give this new medium a try. Some were delighted that the monopoly of the regional ITV companies on television advertising was at last to be challenged. But there was one external threat, an industrial dispute that began innocuously enough at the end of 1982 but grew into a real worry as February approached with no sign of a settlement.

Actors appearing in commercials are paid royalties every time the commercial appears on the screen. The rate of royalty is subject to an agreement negotiated between Equity, the actors' union, and the Institute of Practitioners in Advertising (IPA), the industry employers' group. It is a flat rate dependent simply on the number of times an advertisement is broadcast, taking no account of audience size.

The dispute started over Channel Four, the ITV service aimed at minority audiences that went on the air in November 1982. With viewing figures certain, by design, to be only a fraction of those on the main ITV network, Channel Four would have to charge proportionately lower advertising rates. If advertisers had to pay the same level of royalties to the performers, the sums did not work out. The advertisers would simply not be able to afford Channel Four or TV-am, at least not using commercials featuring professional actors, as most do. Yet Equity would not agree to a royalty that depended on viewing figures. To make new commercials without actors would involve a heavy initial cost and doubtless make for tedious viewing into the bargain.

When it began, nobody believed the dispute would spill over into 1983. The normal pattern in these affairs is for a relatively brief period of unyielding posturing on both sides, followed by a workmanlike negotiation, a few late-night bargaining sessions, the reduction of both sets of irreducible minimum conditions, then a settlement with handshakes all round. There was no reason to think this would be any different.

For weeks it seemed on the brink of settlement. On 17 December 1982, in an interview with Tony Mason of *Campaign* magazine, Peter was optimistic. The dispute, he said, had been an enormous worry until a week earlier. 'It points a pistol straight at the heart of the whole operation. Unlike Channel Four, our revenue does not come by cheque from the IBA. We have to earn it. . . . The ITV companies are losing out from the dispute on what is only 5 per cent of their total revenue whereas in our case it would be 100 per cent. So we would be dead in very short order. I am glad now that the two sides have agreed on a basis for negotiation.'

But relief was not at hand. The following month, less than a fortnight before the on-air date, Peter told David Hewson of *The Times*: 'Our backs are indeed to the wall. As I have made clear over the last several weeks, this dispute and its potential damage to our advertising poses an extremely grave threat to us.' Despite that, 'we are determined to be on air on 1 February. . . . Whether we will still be there on 1 March, 1 April, 1 May and beyond depends on finding a solution to the problem. . . . The financial facts of life would within a matter of weeks destroy the company as an entity. It would cease to exist not because of some whimsical or discretionary decision on our part to quit, but simply you run out of money.'

The board had in fact given serious consideration to delaying the start-up date until the dispute was settled. They were persuaded to go ahead by a virtuoso performance in optimism from Tim Bell of Saatchi and Saatchi, their advertising agents. At the January board meeting he produced wondrous charts and blueprints to show how advertisers would continue to make commercials for screening on TV-am even if the dispute continued, and that any adverse effects could only be a hiccup in the unstoppable surge of the new company's revenue. He thought advertising sales could be as high as £27 million in the first year − £7 million higher than Derek Stevenson's most optimistic forecast and higher even than the figure in the report Timothy Aitken had quoted from in October. Bell closed with a histrionic recitation from Rudyard Kipling's 'If': 'If you can keep your head when all about you are losing theirs. . . .' As one impressed board member put it: 'It was like Laurence Olivier playing an ebullient advertising man to a whole lot of nervous investors.' For the moment, the investors regained their nerve. Most agreed to commit more money to provide a cushion against early setbacks. The show would go on.

Bell had caught them in an upbeat mood. There was genuine excitement at that January meeting of a kind seldom witnessed in dour city boardrooms. This was show business. By the time they were due to meet again the station would be on the air and who knew what heights of critical and public acclaim would have been scaled? It was time for magnanimity, for burying hatchets. That was the atmosphere in which Jonathan Aitken told the board that the sub-committee he was chairing to search for a managing director had decided it would be better to call off the hunt for the moment. Several factors contributed to this decision, prime amongst them the failure of the headhunters to come up with any outstandingly suitable candidate. The interviewing had taken longer than Jonathan had anticipated. Even if an appointment were now made, the new executive could not start work until after broadcasting had begun. The inevitable start-up chaos would linger for some weeks, Jonathan felt, and it would be unfair to pitchfork a new man into it. Then there was the IPA/Equity dispute, in which Peter had fought the company's corner tirelessly and tenaciously. Again, a new managing director would have to come into it half-way. All things considered, it seemed sensible to let Peter stay in both saddles, pedalling the tandem on his own, for a few more miles at least. In proposing that the matter should be shelved until later in the year, Jonathan was gracious enough to pay tribute to the way Peter had nursed the infant company until now, when it was on the verge of taking its first momentous steps into the world outside.

His fellow directors were delighted to agree. The minutes stated: 'The directors decided to record their great appreciation of the achievements of the chairman over the past two years. The enormous volume of work he has undertaken combined with his skills of leadership, eloquence and diplomacy have succeeded in getting TV-am to the starting blocks in record time and good heart.'

So the hatchet was to be buried. Not until much later did it occur to Peter that it might be buried between his shoulder-blades.

The first live test for the new studio and its equipment had come on New Year's Eve, when it was the venue for David Frost's celebrity party, broadcast on Channel Four. Amid the morass of tangled cables, half-finished scenery and all the outward signs of complete panic, the programme went surprisingly well. Peter strolled through it in a kind of daze, clearly astonished that, after the years of planning, it was actually possible to originate TV programmes from this building – a place he could not resist regarding as essentially his own. Even if some of the camera work was less than fluent, the main thing, the astounding thing, the triumphant thing was that it worked.

Next came the dry runs. For three weeks in January the morning programmes were made just as if they were going on air. Real guests were invited and paid real money to be interviewed for the sole benefit of the studio personnel. David and Angela fronted 'Good Morning Britain' for most of the trial period, while Michael Parkinson rehearsed the show he would be hosting on Saturdays and Sundays: he was thought to be quintessentially a weekend person. Anna had taken a week's holiday in the Canary Islands in January with her husband Mark and the baby, but was back for the last two weeks of the dry runs. Deakin was later on to produce that as evidence of an unprofessional attitude but in fact Mark had consulted him in advance about the timing and he had seemed to approve.

Perhaps everyone expected too much in the spirit of euphoric relief that followed the New Year's Eve show: but for whatever reason, the dry runs were generally acknowledged to be terrible. Michael Parkinson's were the worst, although none of it was his fault. One especially dismal morning, when none of his props or his guests or anything else had appeared in the right place and on time, he threw up his hands and temporarily left the studio, observing that he did not need any rehearsal but the production crew surely did.

David was his usual jokey self, but some thought the flipness a bit forced, and he looked drawn. The preparatory period had been exhausting for everyone; and since David had been involved in the project longer than most, he was the most drained. One of the items in the trial programmes was the pilot of a feature series called 'Through the Keyhole', which was to become one of TV-am's most watchable spots. In it Lloyd Grossman, a witty American magazine writer, visits the home of a famous person without being told who lives there. He makes deductions about the owner's identity as he gives a cruel commentary on his or her furniture and knick-knacks. For the pilot, Michael Deakin's London house was chosen. Grossman's remarks about the director of programmes were so vicious and unsparing that Michael was acutely embarrassed. Grossman accused him quite unjustly of being involved in dope smuggling and other nefarious activities. While some of his more unfeeling colleagues were uproariously enjoying the joke, Michael sat crucified by the horror of it. He could see the item contained a germ of a good idea but ruled that in future the victim must be treated more gently, a 'softly, softly' approach employed. He ordered the tape to be wiped clean. Angela was so appalled by the distasteful episode that she took against the 'Through the Keyhole' feature and was never asked to present it.

A few days before the programmes were due to start, David threw still another party. Strictly speaking it was nothing to do with TV-am. It was his mother's birthday and he fêted her at the White Elephant, a night club overlooking the Thames near Vauxhall Bridge. But some of the Famous Five were there, and other guests were surprised by how downcast they seemed. It was in sharp contrast to the upbeat if chaotic mood on New Year's Eve. Since then they had started the dry runs and were depressed at their awfulness. Michael Parkinson was talking glumly of technical deficiencies. For comfort they had the thought that their rehearsal for that IBA interview in 1980 had been a dismal failure, yet the event itself a smashing success. Could they perform the trick a second time, when much more was at stake? Ultimately, they felt sure they could. But it was nerve-racking as they waited to find out.

The BBC's 'Breakfast Time' show was born at 6 a.m. on 18 January 1983, and the first thing everyone noticed about it was Frank Bough's sweater. Bough is an old BBC hand, formerly host of the Saturday afternoon sports programme, with receding hair and a comfortable, safe voice. He suggests armchairs, spaniels and well-worn carpet slippers. You expect to see him produce a pipe from his pocket and knock it out on an ashtray. The sweater was just the right touch. It fitted in with the relaxed yet workmanlike mood that Ron Neil, the editor of the new programme, had worked hard to create. The other presenter, Selina Scott, was a perfect foil – a vivacious and striking former news-reader in her twenties, with an endearing habit of fluffing her lines just enough for viewers to notice but not to interrupt the flow of the programme.

In the contest for morning viewers, the BBC enjoyed a number of advantages over TV-am, apart from their two-week start. With the advent of the new service, their old Lime Grove studios in West London were renovated to create a round-the-clock topical programmes studio that would act as a base for 'Breakfast

Time', 'Nationwide' in the early evening (now renamed 'Sixty Minutes'), and 'Newsnight' on BBC-2 late at night. This allowed them significant economies of scale and made it worthwhile to introduce an advanced newsroom computer to service all three programmes. Its cost of £500,000 could be split between the three budgets. TV-am had a comparable machine that they had to pay for themselves. The computer allowed the programme staff to keep track of news stories and other items as they developed. It made for slickness of presentation. In conjunction with this came an advanced computer graphics unit, costing £150,000, introducing techniques never seen before on television. The result was a show that looked as fresh and bright as people were hoping to feel at the start of the day.

Fresh and bright but undemanding — that was the formula Ron Neil and his team arrived at. They agreed with Peter Jay that they were aiming for a tabloid newspaper audience — or in BBC terms a Radio 2 rather than a Radio 4 listener. They sensed that the Radio 4 person, the *Daily Telegraph* reader, was likely to stay loyal to the popular 'Today' show and its presenters John Timpson and Brian Redhead. Those most likely to sample the novelty of TV in the morning would be of a less serious cast of mind. So along with service features like news, the weather and road conditions, they ran the plumpish, campish Russell Grant to provide a daily horoscope and banter with Selina. Diana Moran, a lithe and enthusiastic keep-fit expert who quickly became known as the Green Goddess, surprised people in public places and made them perform physical jerks. The editor of the William Hickey column on the *Daily Express* provided a weekly roundup of society gossip. As for news, they had the inestimable advantage of their own extensive and smoothly-running news service, while TV-am were starting theirs from scratch. Even the BBC, however, cannot regulate the supply of newsworthy events. It was not their fault that the main story on that first day, an air crash in Turkey, scarcely fitted the companionable mood of the rest of the show.

Viewers liked the BBC product. For the first few days the rating for the peak quarter hour — a controversial refinement of audience measuring techniques introduced specifically for breakfast television — approached two million, or 8 per cent of households. That was double the rating TV-am had been banking on – and calculating advertising rates by – for their own programme. It inspired optimism in Camden Lock. If the audience was there, surely there was no reason why they should not gain at least a half share of it, and with any luck more.

The press were enthusiastic, too. There were specific criticisms but the overall view was that, given the multiple pitfalls inherent in any new project, it was a creditable beginning. But the pundits seemed to be holding something in reserve. They did not want to squander their stocks of the most extravagant adjectives on the BBC's effort which, however worthy, they felt was bound to lack the professional gloss of the Famous Five's rival product.

Newspaper interest in the five was tremendous, almost obsessive. The press loves nothing better than star names, especially star names with photogenic faces, like Anna's and Angela's. Between them, the five were interviewed for almost every conceivable paper. The tone was consistently upbeat. They were too professional to let the disappointment of the pilot shows filter through to the public. David, for example, was asked by Bryan Appleyard of *The Times*, in the week before the debut, whether the Famous Five might not be a trifle old hat as celebrities for a station seeking to locate itself on a new frontier.

'It's one of the things that's come out of the pilots – they're terrific,' Frost replied. 'Parky is doing weekends and he's so right for that. Robert and Angela are great whether they're doing the news or the 'Good Morning Britain' show, and so is Anna. ... One thanks God we made the choices we did as far as people are concerned.' Michael Parkinson told the *Sunday Express* he had never been more excited by a new venture.

If editors and proprietors ever had any pangs of worry about giving such intensive publicity to what was in effect a rival concern, competing with newspapers for audiences and advertising, they overcame them. Fleet Street's only concept of competition is with itself, with other newspapers. Experience shows that stories about television, with star names, sell papers. So if the *Daily Mail* secured advance pictures of TV-am's keep-fit lady, their answer to the Green Goddess, the *Express* would hit back with an exclusive account of what Anna Ford and David

Frost would be wearing for their first show. Much was made of the heroic stunts involved in making the opening title sequence for 'Good Morning Britain'. The idea was that groups of people and birds should spell out the title on various unlikely locations. Four sky divers held a 'Good Morning Britain' banner aloft, sailors formed the words on the deck of the aircraft-carrier HMS Hermes, hundreds of people were taken to Bristol Downs to do the same thing there. Most ingenious of all was to enrol the pigeons of Trafalgar Square. Food was placed on the ground in a pattern forming the letters of the three words. When it was uncovered, the pigeons swooped down and obligingly posed for the aerial cameras. It all made a tremendous photo-story for the papers.

By the number of column inches devoted to TV-am, an innocent visitor might have judged that a major revolution in British life was in the making, rather than a simple extension of television broadcasting hours. By the time 1 February dawned, the press had managed to whip itself up into a state of slobbering excitement – even if, as initial audience figures would show, the general public were better able to resist the hype and contain their anticipation.

The atmosphere in the studio itself was a frenzied mixture of euphoria and deep exhaustion. Both qualities were reflected in a letter, signed by Peter and David, that was distributed to the staff on Monday, 31 January, the eve of going on air. 'Tomorrow, Tuesday, is D-Day,' it began.

> Your enthusiasm and hard work – particularly in these last two weeks – have been heart-warming to behold and have been the key to the fact that we are now as ready as we shall ever be to burst successfully on to the world. I know it has been fair hell and average chaos. . . . Even if we have to open the TV-am home for exhausted operatives it will have been a great thing to have launched TV-am, knocked hell out of the competition and created the most valuable addition to the British way of life since Yorkshire pudding (no allusion to the Programme and Operations Department intended). . . . Good luck and keep paddling.

Michael Deakin's letter to his staff was more concise: 'Good luck for tomorrow – we all need it.'

A few newspaper reporters were given a free run of the studio
on the Monday to watch the last twenty-four hours of
preparation for the first programme. Some of the senior staff
had been against such blatant exhibitionism but Peter overruled
them. A vivid description, reflecting the sense of anticipation,
chaos and sheer terror rife at Camden Lock, was written by
David Hewson in *The Times*. As he reported, the muses who
regulate the flow of news, having failed to produce a dazzling
story for the launch of the BBC's 'Breakfast Time', proved their
even-handedness by providing nothing special today, either, for
a news-reactive station to react to. The biggest story was the
water workers' strike, and that had been running for days. The
pivot of 'Good Morning Britain' was to be David's long
interview with Norman Tebbit, Secretary of State for
Employment. The strike would presumably be raised in that
but the main subject was to be unemployment. An opinion poll
conducted by TV-am would be brought in by David as evidence
of the public's deep concern.

That was the first item discussed at the 9.30 a.m. conference:
all very sober and responsible. But, as Hewson reported:
'Weighty matters dismissed, TV-am's perpetual struggle
between lofty current affairs ideals and the need to grab
audiences – and consequently advertising – swiftly surfaces.'
The issue concerned the station's weather forecaster,
Commander David Philpott. He was in many respects a curious
choice for the post, a retired naval officer who had simply written
in to apply when he read the *Guardian* advertisement, like many
others without obviously relevant experience. It is not hard to
understand how Michael Deakin and the others came to think
of him as an ideal weather man. The weather and the sea have
a natural affinity. He would stand by his charts barking his
predictions like a captain addressing a mess full of orderlies.
Then David could indulge in nautical or military banter with
him. It had a delicious touch of camp.

Whether the Commander would stand up or sit down before
the camera was the issue now to be debated at the conference.
It ought to have been decided days earlier but it remained
unresolved. Some thought that, to distinguish him from other
weather forecasters, he should be seated. They were overruled

and it was agreed he should stand, although not necessarily to attention.

The next decision was whether David Basnett of the General and Municipal Workers Union should be invited to give the union's response to Tebbit. It was agreed first to invite him, then not to and finally, late in the day, to invite him again. The decision who should be the chief show-business guest was again left until the last minute, reflecting Hilary's view that, even if it would be less damaging to the nervous system to arrange these things well in advance, you keep a sense of immediacy by delaying it until it is almost too late. After all, few people have other commitments at 7 a.m. They decided to invite the comedian John Cleese, who turned up in his pyjamas. But two other last-minute invitees, motor cyclist Barry Sheene and ex-boxer Henry Cooper, turned down the chance to discuss male grooming and make history. Before the meeting someone looked in the door and said: 'I'm going to assault the chairman with a catalogue of complaints – anybody want to add some?' Fair hell and average chaos.

There was no shortage of complaints. Hewson reported how, 'starved of breaking news stories, the night staff start to moan about the long hours, the faulty technical equipment and, more than anything else, the conflict between the rich promise of a new form of intelligent popular TV journalism promised by the consortium when it won its franchise and the reality – a much more down-market version.' He quoted a new recruit as complaining: 'The trouble with the management is that they couldn't find their backsides with both hands. Creative tension isn't the half of it. This job is three years of high intensity followed by immediate professional burnout.'

Robert Kee and Angela Rippon were scheduled to present the 6 a.m. 'Daybreak' programme. This was an expedient for the first few days only, in order that all the Famous Five should establish their screen presence right away. Later, according to the roster, 'Daybreak' would be taken over by two younger (and earlier-rising) presenters, Linda Berry and Gavin Scott. Robert arrived in the studio at about 2 a.m. and Angela not long afterwards. There was indeed not much news and what there was became tedious after the third or fourth quarter-hourly repetition. But Angela and Robert handled it professionally, perched a bit awkwardly in the news room, where people would wander on and off screen among the desks in the background. This was an idea Michael Deakin had picked up from America, where it is supposed to suggest the immediacy of breaking news – hard to sustain when no news is breaking.

It was an unexceptional hour of news, a little thin on action film and on original stories, but competently put together. There was precious little evidence of the radical new approach to news that had been promised, but this was the opening day and it would have been churlish to make too much of that. What everyone was waiting for was seven o'clock and the start of 'Good Morning Britain', the main feature. The pigeons flew into place and David said: 'Hello, good morning and welcome.' Apart from the overlong interview with Tebbit, who had little of significance to tell David and the waiting millions, it went well enough. David and Anna sat demurely on the edge of their sofa, in a setting designed as a ranch-style living room, the one bouncing with confidence and the other looking only slightly overawed. Those viewers who had drunk in that flood of advance publicity must have felt it a privilege to be allowed to watch two such exceptionally celebrated people. John Cleese not only appeared in his pyjamas but also lay on the floor for a while. Commander Philpott stood straight as a ramrod and read the weather forecast with the confident air of a man who has braved worse storms. The first public airing of 'Through the Keyhole' was targeted on the house of Stirling Moss, the former racing driver.

Viewers might or might not have been taking all this in but in the studio few people were doing so. Peter had invited a number of guests, mainly from the world of television, to watch the first programmes being broadcast. They included Lord Thomson, chairman of the IBA. As will often happen, the guests took to chatting among themselves rather than watching the action. When it was over the Famous Five went upstairs to join the others. Only one thing was left for them to do: they had another party.

CHAPTER 5

Too Many Friends

The newspaper critics next morning were nearly unanimous. This was it, this was the real thing. The BBC's effort, quite worthy in its way, was now seen as nothing but a warm-up for the main event. Here was Herbert Kretzmer in the *Daily Mail*, under the headline MY MONEY GOES ON THE JAY ALL-STARS:

'It may be rash to predict a technical knock-out during the first round of a big fight, but my money's on Peter Jay's new TV-am to score a significant ratings victory in the weeks and months ahead.

'The new service launched itself into the cold dawn sky yesterday with the jaunty and confident air of one which will brook no rivals.'

It was, wrote Kretzmer, 'a sparkly, well-orchestrated debut'. He went on: 'The public has always been a sucker for a famous face and TV-am has lots of them. . . . Faced with the choice between the paternal ease of Frank Bough on the one side and the Jay All-Stars on the other, the great British public may be widely depended upon to go for the glamour. It is glamour that will almost certainly gain the glittering prizes.'

Warming to his theme, Kretzmer went on to note the show's 'brisk sophistication'; but he did sound a note of caution. Compared with the BBC's array of astrologers, gossip writers and guests from show business, 'Good Morning Britain' was 'unexpectedly up-market, which may well work against it in the long run.'

No such doubts afflicted Hilary Kingsley of the *Daily Mirror* who, under the headline TV-AM LOOKS A WINNER, described the Famous Five sailing 'calmly into action like the polished professionals they are'. In her *Daily Express* review, headed IT'S THE TOP OF THE MORNING SHOW, Jenny Rees declared: 'This is a bit more like it. . . . To me, they've got the show pitched at exactly the right level for the morning.'

The one dissenting voice at the popular end of Fleet Street was also to be heard from the *Daily Express*, in the person of their controversial and frequently bitchy columnist Jean Rook. She did not like the Famous Five one little bit and was not afraid to say so.

'What I couldn't swallow,' she wrote, 'were the hard-boiled, cynical, snide, patronising people with whom I was supposed to enjoy breakfast. . . . Give me bacon with Bough and sausages with Selina. They may in their darkest dawns look half-dressed and half awake but at least they're the sort of happy family you'd want to have breakfast with.' Miss Rook found only one redeeming feature. Like the other critics, she was bowled over by the weatherman, Commander Philpott.

At the *Daily Telegraph*, critic Sean Day-Lewis avoided the invidiousness of choosing between the rival services by writing: 'a plague on both your houses'. But he did observe astutely that on the evidence of the first day TV-am's news effort was not a patch on the BBC's.

Some of the 'heavy' newspapers were less enthusiastic than the tabloids but the balance of the comment was greatly in favour of the new station. Everyone at TV-am was properly pleased, although there were none of the extremes of jubilation that had followed the gaining of the franchise. They did not really have time for that, with three and a quarter hours of live programming to originate every day – much more, as Peter never tired of pointing out, than any other commercial station was committed to. And the operational difficulties that had become apparent during the dry runs did not go away. There was still pressure on the editing machines and long waits for their use. The shift system meant that some people were alarmingly overworked while others were earning generous 'unsocial hours' payments for sitting around doing nothing.

The presenters, who had enjoyed less of their prized 'input' than they had been led to expect, now found that they had virtually none at all. It was physically impossible for them to stay for the afternoon and evening planning meetings after their early start. Items were therefore sprung on them when they arrived in the morning: that was how the female circumcision and G-spot features had slipped under Anna's guard. Some were sprung even later than that, only minutes before the show went on the air. The programme staff were supposed to give the presenters detailed briefing documents about the items in the show. If they did not arrive until the last minute the presenter had no time to adapt them to his or her own idiom, and was obliged simply to read out what the researcher had written. This robbed the presenters of the individuality that was supposed to be their strength.

Almost the worst failing, one that gave the programmes an irredeemable sense of amateurishness, was that they seldom ran to schedule. The *TV Times*, the ITV programme magazine, printed outline schedules and these were picked up in the daily papers. Items such as the weather, the review of the morning press, the shopping news, were all supposed to appear at fixed times but in the early days they seldom did. One function of a morning programme, on either television or radio, is to set clocks – or body clocks – by. Viewers should know that if the sports news is on, it is time for them to be wiping that last bit of egg from their mouth and shouting at the children to get ready for school. If they are still at home by the time the weather forecast comes round, they will have to run all the way to the station to catch the train.

Over at the BBC Ron Neil was acutely conscious of this and made sure everything came up as advertised. He designed the programme like a cake, which could be sliced into meticulously calculated chunks. Except on those rare occasions when there was a remarkably strongrunning story, Neil was rigorous in ensuring a disciplined adherence to schedule.

Back at TV-am, Hilary Lawson's programme was not so much a neatly sliced cake as a loaf cut up by a clumsy child with a blunt penknife. The size of the slices was irregular and unpredictable. What should have been a dainty, wafer-thin sandwich slice came up at misshapen doorstep thickness. People switching on for the press review at, say, 7.56, would instead see the tail end of David's interview with a man who had bred a miniature rhinoceros. Then the news would intervene and the press review would turn up some fifteen minutes later, when others were switching on for the traffic report. It was not all Hilary's fault. In particular it went back to the dispute over presenter power. The editors found that their instructions to some of the presenters to wind up interviews on time were quite often ignored.

A few editors attributed this to stubbornness on the part of the presenters, peeved at having less influence than they thought on programme content and seeking to put their stamp on the shows like this. But it was not that. In David's case it sprang from his deeply-held commitment to free-flowing, spontaneous television that creates its own pace and is not susceptible to detailed advance planning. His most successful past programmes had been based on this practice. It works well late at night, when people are relaxed and unhurried, but looks sloppy in the morning as viewers are struggling for a reasonably orderly start to the day. David did not appreciate this distinction and, if he felt an interview was going exceptionally well, he would ignore a director's instruction to wind it up. He would let it run for a bit. Anna did the same thing once or twice: in her case the reason was not her programme philosophy but simply her limited interviewing experience. The hardest part of an interview can be the ending of it, cutting off a garrulous guest in full flow without making the presenter or the guest look foolish.

The problem persisted. As late as March John Holme, an editor-of-the-day, was seen after a show exchanging angry words with Angela on that very topic. Derek Stevenson's advertising department found it especially irritating, for if interviews were allowed to overrun it meant that other items simply had to be left out. Some advertisements are sold on the understanding that they will be screened immediately before or after specific segments in the programme. If the segments did not appear as planned, or if they were radically retimed, this undertaking was impossible to fulfil.

The satisfaction over the initial press reaction was short-lived. Although official audience ratings of television programmes do not appear until the end of the week, it is possible to get a reasonably accurate idea of how you are doing from the 'overnights', daily ratings based on a smaller sample. It was soon clear that the Famous Five were not getting anything like the audience they had hoped and were being badly beaten by the BBC. The final figures showed that in the first week – when curiosity, fanned by the excesses of the press, should have been at its height – the BBC were getting twice as many viewers as TV-am, 1,600,000 as against 800,000. It was alarming because 800,000 was the minimum figure required to sustain the advertising rate Derek Stevenson had set. If it fell below that – and the trend from the very first day was downwards – the rate would have to be reduced and the prospect of profitability would fade.

More detailed figures, when they became available, were still more worrying. They showed that even on the very first morning, despite the advance publicity, the BBC outscored TV-am by a margin of 56 to 44 in the London area and by 72 to 28 in the other sample area, Lancashire. By the first Friday those figures had reached 75 to 25 in London and 83 to 17 in Lancashire.

At the end of the first week Peter received a phone call from his friend John Birt. Quite inadvertently, Birt found himself playing the role of Cassandra in this saga; or, on a more down-to-earth level, he was like a football supporter trying to egg his team on with repeated cries of 'What a load of rubbish!' It was his stern coaching before the IBA interview that had done as much as anything to win TV-am the franchise. Encouraged by that clear success, he decided it was time to warn Peter about the poor quality of his programme, not precisely a load of rubbish, but with some grave and possibly fatal failings.

Birt had been at the launching party and as a professional he did not share in the first-day enthusiasm. He told Peter he had problems. Throughout the first week he detected still more flaws and, he thought, growing disorganisation. So he invited Peter to drop in at his South Bank office for a drink on his way home. He realised that anyone of a more resentful cast of mind might react against gratuitous advice, but none the less he felt he owed it to him as a friend – and one of far greater experience in television – to speak his mind.

It was a gloomy encounter. Birt told Peter he had failed to work out relationships between the presenters and the production staff, and even between the presenters themselves. David and Anna seemed not to form any kind of coherent team. Insufficient thought had been given to precisely what audience they were aiming at – in contrast to the careful and correct aim of the BBC. In any case, there seemed no structure to put a programme philosophy into effect. It seemed to him that the new station lacked the touch of a strong and imaginative programme executive. The presenters, however talented and attractive, could not make poor programmes acceptable. Birt said he had not expected that the mix would necessarily work right away: but he saw no indication of any movement in the right direction.

The second week's audience figures were so alarmingly lower that Peter could see some kind of emergency action was needed. The daily average was down to 500,000 – well below an economic level – and the BBC had climbed to 1,800,000. Now an extra problem, low morale, was added to all the others. Peter did two things. He commissioned some audience research to determine just why people were not responding to the programme. And he formed an action committee to meet in his office at ten every morning. Consisting of executives just below the top level, its role was to follow through the most serious operational snags that were occurring and discover their root cause. Despite that, the snags did not go away. Michael Deakin summoned frequent crisis meetings of the programme staff. 'It was a cycle of disasters and pep talks and more disasters and more pep talks,' said one. Ratings for the third week held fast at half a million but in the last week of February they dipped disastrously to 300,000. Even that was misleadingly high because the ratings are calculated on the number of viewers in the peak quarter-hour. TV-am's peak was not being reached until after 9 o'clock, when the BBC went off the air. In the first hour of transmission after 6 o'clock the TV-am audience was below 100,000 and thus earned a zero rating. Only Michael Parkinson's weekend programmes, without BBC competition, were getting respectable figures.

The press is a fickle ally. Newspapers that less than a month earlier had been giving the new station yards of free publicity were now playing their role in the snowball effect of bad ratings. Headlines such as FAMOUS FIVE FLOP persuaded the public that TV-am was not worth spending their time on. So the ratings deteriorated still further and, as the headlines grew larger and gloomier, the excuses offered became increasingly desperate. It was said the ratings were wrong. Then it was maintained that the viewers were switching off when advertisements were screened – an oddly self-destructive argument for a commercial station to deploy.

On 20 February the *Sunday Times* reported that the five were 'getting a pasting' from their rivals at the BBC, and were being forced almost to give advertising time away. On their original estimate of viewers, the sales department had been asking more than £2,000 for a 30-second slot. Now they were promoting a bargain offer of £600 for the same half minute. Alan Prince of the Zetland advertising agency said that a 30-second slot was 'not worth a penny'. While he would pay £1,000 for a '3' rating (600,000 viewers, or 3 per cent of the potential audience), a zero rating was not worth anything to him. The *Sunday Times* sought the views of other advertising people, who were blunt in their criticism of the programmes and the presenters. One, commenting on David Frost's infrequent appearances on British television in recent years, called him 'an unawakened Rip van Winkle' and declared that both he and the shows were out of date. Said another: 'He's very sharp but people don't want cleverness early in the morning.' Another view was that they were all 'too polished'.

By now the investors were getting alarmed. Teething troubles were one thing, but they were faced with a virtual disappearance of the company's revenue base less than a month after going on air. Advertising was being sold at rates that were a joke, though a poor one. On the surface it did not look too bad. Revenue for February was reported to the board as being £1 million, with £700,000 booked for March. This was below break-even but it was a fair performance for a launch. The reality was that all this advertising had been booked before 1 February. Since the station had been on the air, scarcely any new business had come in. To meet their revenue forecast, the sales department should be gaining new bookings at a rate of at least £300,000 a week. The actual rate was around £50,000.

Just before the station went on air, Peter had arranged a refinancing of the company in the face of the increased startup budget. The new figures suggested a need for £15.7 million capital in the first year. The original funding, including the £2 million standby arrangement with the existing shareholders, amounted to only £10 million. The shareholders, including the presenters and founders, were offered extra shares at £3.50 each, at a rate of three for every four held. Not all the institutions took their full entitlement. Eastern Counties Newspapers and Robert Stigwood took part of it while Paul Hamlyn's Octopus group took none at all. Most of the founders and presenters bought the new shares, but Anna did not. Even with an additional bank loan facility of £2.75 million, this left a shortfall of £1.1 million. It was met by bringing in three new investors: the Prudential Assurance Co., the Merchant Navy Officers Pension Fund and the Alva Investment Trust. Peter's buoyant mood as the station began transmitting was indicated by the letter he wrote to shareholders to announce these moves and to invite them to approve them at an Extraordinary General Meeting on 7 February: 'I look forward to seeing you then, not least to celebrate my birthday.'

Now, only three weeks later, the shareholders had nothing to celebrate. Without a quick improvement, and despite the new financing, the company would be insolvent before another month had passed. Their investments would disappear with it. Peter, who had a significant portion of his own savings tied up in the venture, knew he could expect trouble from the board. By the end of the second week John Birt was not the only person cautioning him, as friend to friend, that he had to do something decisive.

On Thursday, 17 February, the results of the research Peter had commissioned came to hand. They were broadly similar to the reactions *Sunday Times* reporters had gleaned from advertising agents. The show was difficult, remote and too tough to cope with at that hour of the morning. That weekend Peter, Michael, Hilary and the presenters met. Although they had serious doubts about the validity of the ratings, and had

141

expressed them publicly, they agreed on a new look for the programmes, to come into effect on Monday, 28 February. That was the start of the station's fifth week, when in any case the roster was due to turn and Anna and Angela would share the presentation duties. Peter and the presenters believed that this unique all-woman team were the ideal crew for the relaunch. Michael Deakin and Hilary had their doubts. They feared the pair would combine as ineffectively as oil and water. In particular, Michael had his doubts about Anna. He seemed to be in a minority of one on the Anna issue, though he had nothing against her personally. After all, he had spent Christmas enjoyably enough at her home the previous winter, and had been touched by the trouble she had taken to make things pleasant. She was undeniably attractive and personable, if a little trendy. It was just that he did not share the high opinion of her intellectual prowess held by Peter and the others, who appeared to think of her as the Brain of Brentford, a combination of Wittgenstein and Helen of Troy. Michael saw no evidence to support that. But above all, he thought breakfast television was not her strong suit.

Yet he was disinclined to argue and it was agreed that the two women should be promoted heavily on the ITV network in the week before their debut. Other changes were made to the programmes, all with the object of making them more accessible. The news-orientated 'Daybreak' hour was to be halved, meaning that the freshly-leavened 'Good Morning Britain' would start half an hour earlier, at 6.30. A quiz would be introduced with a portable television set as the daily prize. Individual items in the show would be made shorter, there would be more features and more guests. Peter, his optimism still largely intact, described the changes as 'swift, selective and sensible'. What he did not admit was that the mission to explain, insofar as it had ever been launched, was now effectively grounded.

The February board meeting was not held until 3 March and things did not go as badly for Peter as they might have done. The directors were supportive. It was their first meeting since the station went on air and although events since their ebullient January session had been disquieting, they still felt it incumbent on them to congratulate Peter for having launched the station. He was able to tell them about the programme's new look, about how hopeful he was that it would succeed. The directors assured him they would not panic. They said they would await the results of the changes with interest, which was an understatement. Michael Deakin and Geoff Smith (now on the board) said part of their problem was a desperate shortage of staff. The directors, led by Jonathan, urged Peter not to be shy of spending more money if it was needed to get things right. That ubiquitous ha'porth of tar was mentioned again. Peter said he was still determined to clamp down on spending as tightly as possible. He looked forward to bearing hopeful news at the next meeting.

Already, however, events had been set in motion that would deny him this wish. On Monday, 28 February, the first day of the new look, Anna and Angela duly appeared. To Michael, they did not seem much preferable to Anna and David. Derek Stevenson, who for weeks had been forced to listen to scathing criticisms of the programmes from his advertising customers, agreed with him. Though superficially friendly and generous with each other, the two women somehow could not overcome the subliminal impression that they were in competition, trying to score points. They seemed to vie with each other in smugness. Their chat was of the demure, cautious quality found at suave West London dinner parties. It was with mixed feelings, then, that Michael took a call from Anna late on Monday afternoon. She said she had 'flu and would not be able to appear the next day.

David was keen to stand in for her. Aware that people were blaming the disaster of the first month partly on him, he was anxious to show that, without Anna's unsettling presence, he could perform well. Besides, the press, in their obtuse way, had persisted in interpreting the Monday switch of presenters as meaning David had been axed. Although Howell James, the press officer, insisted time and again that it was all part of the master plan drawn up nearly a year ago, they would not take that on board. Michael had reservations about David. Derek Stevenson, after the anti-Frost flak he had been getting from his advertisers, was aghast. What clinched it for David was his assurance that he could only do it for one day, since he was flying to America on Wednesday. If Anna's illness lasted longer, someone else could be found. David wanted it badly and it was only this once, so Michael and Peter agreed.

Next day Anna phoned to say she felt worse and might be away for the whole week. Michael Parkinson was asked to stand in for her. He and Angela were a perfect partnership. Everyone agreed that the shows on the Wednesday, Thursday and Friday of that week were the best in TV-am's brief history. Michael Parkinson exudes an air of unhurried confidence and this rubbed off on Angela. They were two people you might actually enjoy having breakfast with. Derek Stevenson had the rare pleasure of receiving congratulatory calls from his clients.

Anna, recuperating now, told Michael Deakin she would be back the following Monday, 7 March. His reaction was: 'Why? Why should I be forced to take her back when I'd really prefer to stick with Parkinson?' He reflected ruefully that in a conventionally constructed TV company it would be a simple matter to suggest to Anna that her illness might be more serious than she thought and she should stay at home for a few more weeks until she was really better – and until the roster had safely removed the threat of her reappearance in front of the cameras for a while. But this was not a normal company. Anna was one of Peter's precious assets and only Peter could give her unchallengeable instructions. Before going to see Peter, Michael spoke to Derek, who agreed to support him by writing a memo to Peter saying how much better Parkinson was than Anna in terms of his sales.

Listening to what Deakin had to say, Peter could sense that this was one of the trickiest problems he had to solve at TV-am. In his pragmatic way, he tried to isolate its essential elements. First there was the human factor. Anna would clearly be mortified if she were told to stand down. She did not accept that the first disappointing month had been her fault and she was anxious to prove that, without the jumpy David as her partner, she could pull in the viewers. She and the other presenters had discussed the low ratings with Peter. They insisted that the fault was not theirs but that of the programme staff, Michael's and Hilary's people, for providing inadequate backup. He would obviously prefer to avoid provoking Anna if he could; but the other human factor concerned Deakin. Peter knew he was sensitive about presenter power and he saw his point. Certainly Peter did not want to be seen taking sides with the presenters against his chief lieutenant. His schooling at Winchester had given him too formal a sense of hierarchy for that. The head must back his prefects.

But this was not Winchester, it was a company with large amounts of other people's money at risk. What he must do was ignore the human factors so far as he could and be guided by the facts. The trouble was that there *were* no hard facts in the case. Anna and Angela had been on the air together for only one morning, so it was quite impossible to say whether they were a winning team in terms of ratings. Subjectively, most people seemed to agree that Michael Parkinson and Angela were a more successful pair but objectively the only evidence was the overnight ratings for Wednesday and Thursday. They did in fact show a slight improvement on the previous week, but it was very slight and, in his view, inconclusive, balanced by the fact that Anna and Angela had been promoted heavily in advertisements. It would make everyone look foolish if they were to split the team after only one performance. And although he was sure Deakin was doing only what he thought best for the company he did recall an incident a few weeks earlier when Deakin criticised Anna's performance privately. Peter felt obliged to warn him not to let any hint of that criticism spread through the building. It would be a self-inflicted wound on the company. But Peter trusted Michael and was sure his motives were not at heart personal. He told him to phone Anna tactfully and see how she felt. If she would agree to the substitution, then the human factor would largely be eliminated and Peter could, with an easier mind, go along with Michael's plan. But he doubted whether it would be that simple.

On the phone, Michael thought he handled Anna rather well. He put it to her the reason he wanted to stick with Michael Parkinson was that the programme now needed above all a measure of stability. 'I just can't take any more changes, love,' he complained. Anna listened carefully and said little. She was not going to be hurried into an instant response. She would think about it and call back.

When she did, her answer was decisive. She would have no truck with the plan to replace her. She was not going to be 'bumped' in what she saw as a panic reaction to low ratings. To agree would be to accept by implication that she had performed poorly, and what effect would that have on her career? She had sacrificed a great deal for this company. She had spent two years off the air – two of what could have been the most productive years of her life. She seemed hurt and angry. 'Ungentle sir,' she told Michael, 'I will do no such thing.'

Ungentle sir – Michael liked that. He wrote it down among the growing volume of notes he was making on the developing crisis. He pondered what to do next. He had received a clear impression from Anna that if he insisted on changing the roster she would consider leaving the company, and that at least one other presenter might go in sympathy. It was now patently a matter for the chairman, even if it had not, in Michael's view, been any of his business before. He went to see Peter and Peter agreed to visit Anna in Brentford. As it happened, she lived not too far from his own house at Ealing.

Anna was calmer by the time he arrived. They were, after all, good friends. Peter, too, was soothing. He explained that it was perfectly reasonable for a company to look at various options and discuss them with colleagues. Anna accepted that, but was not prepared to allow that one of the options should be her head on the block, her professional destruction. Peter tried to assure her there was no question of that, whatever happened. But she was adamant. Towards the end of the conversation, her husband Mark Boxer came in to lend his support. Mark suggested there should be three presenters next week – the two women and Parkinson – but Peter did not seem to be listening.

When he reached home Peter telephoned Deakin and reported on his meeting. They would now have to decide what was best for the company. To Peter, it seemed evenly balanced. He understood Deakin's reasons for wanting the change but another consideration was the effect on staff morale if the pre-arranged roster were altered now. The press would represent it as yet another instance of abject panic and that in turn would harm advertising sales. Michael agreed it was difficult. He conceded that Peter should make the final ruling. Peter decided that Anna should return to the screen on Monday. Michael concurred.

Although he had, like a good soldier, gone along with Peter's decision, Michael saw this as crucial confirmation that he had no power to control the presenters. If he could not deploy the presenters as he liked, how could he properly shape the programmes of which they were the pivots? His sensitive antennae had picked up the criticisms being made inside the building of his and Hilary's performance. But surely it was quite unfair to blame them when, as the Anna incident demonstrated, attempts they made to improve matters were resisted. He felt like the captain of an English cricket team, blamed for poor results even though the selectors, against his advice, had failed to pick the best players. Maybe, before things went irredeemably wrong, he could appeal even over the heads of the selectors to the board of control. He thought about it for a couple of days and then, on Sunday, 6 March, he sat down and wrote to his long-time confidant, Jonathan Aitken.

What later came to be known as the 'smoking gun letter' was, on the face of it, an innocuous document. Michael first gave a factual description of the Anna incident and then expressed his fear that he was losing all control over the presenters. Peter, though, did not interpret it like that. He told Michael, who showed him the letter a few days later, that if he thought it innocuous he was being naïve. To any outside person it would read as an account of a wrong decision, weakly made under pressure from one of the presenters. Michael insisted that had not been its purpose. Hadn't he, after all, stressed in the letter that he had agreed with Peter's decision over Anna? The motive for writing it, he said, was to place on record the limitations of his power, so that when the time came to apportion blame for the low ratings, it would not fall on him. And the reason he, with his known distaste for writing things down, had chosen a letter rather than a verbal *démarche* was, he told Peter, chiefly to spare his feelings. Michael was conscious of his reputation as an anecdotalist. Some of his colleagues on the board felt he had gone too far at the September meeting, when he had complained of Peter's aloof management style while Peter was out of the room. He agreed that he may have said more than he really intended for the sake of a good story and he felt guilty about it. This time he would write it down, to avoid any such possibility.

In the mythology surrounding the power struggle at TV-am, rather too much may have been made of this letter. It makes a neat conspiracy theory: executive feels his job at risk so he writes to his old friend on the board criticizing the boss. But the letter was not the prime cause of the process that Jonathan, when he read it, proceeded to set in motion. At most it triggered an explosion that was always likely to occur. Jonathan would later describe it to friends as a tinder brought to a pile of wood. While it was, in a strictly logical sense, the cause of the subsequent conflagration, the wood was by then so parched that a stray spark might have set it ablaze at any time.

The timing of the letter was all-important. For on the weekend that Michael wrote it the worst-ever viewing figures for TV-am had just been confirmed. During the last week in February their peak quarter-hour rating had, as the additional figures had indicated, fallen to 300,000, less than a fifth of the audience for 'Breakfast Time'. Michael knew this made him vulnerable and so did Peter. The newspaper critics were writing what were in effect obituaries for TV-am just five weeks after it had been born. The view was gaining ground that tinkering with the presenters' roster was not enough: the whole concept had to be scrapped and thought out again from scratch.

Peter felt it was time for a morale-raising gesture. On Tuesday, 8 March, before he had been told about Michael's letter, he summoned the staff for a meeting, where he delivered a fighting speech. He pointed out that the new figures referred to the week before the 'new look' of 27 February had gone into effect. He said the institutional investors continued to support him and the company. 'We're on the attack,' he told the worried workers. 'We've just got to hang in there and keep our nerve.' And he added, with a nicer irony than he knew, that the company's future was safe for a hundred years. 'A station called TV-am will still be around then but whether we're in charge of it is another matter.'

In a television interview he was equally buoyant. 'We have some weeks and months to get it right. It does not help to make panicky, silly and ill-considered changes. The whole atmosphere at TV-am is one of attack.' Michael Parkinson told reporters that he had been through all this before at London Weekend. 'We will get it right but it will take two or three months.' Michael Deakin commented: 'It is just going to be a long haul.'

Jonathan Aitken read reports of those statements on Wednesday – the same day he received Deakin's letter, for he had been out of the country at the start of the week. It seemed to him, judging from both the viewing figures and the letter, that they simply could not afford to wait the weeks and months Peter and the two Michaels were talking about. The letter brought back all those doubts that had been accumulating about Peter's stewardship. The only real test of a manager, he believed, came at a time of crisis. Here was Peter's first real crisis, one so grave that it posed an imminent threat to the company's survival. It seemed to Jonathan that, by his shilly-shallying over Anna, Peter had shown himself unable to take the decisive action required. His 'swift, selective and sensible' changes of the previous week had been little more than window dressing. Sterner measures were needed. The whole 'Famous Five' approach seemed to Jonathan – and to a growing number of press critics – to have been discredited. So had the production staff. Swift surgery was almost certainly required in both areas, even if it did strike at the egos of the presenters and even at his chum Michael. The Anna incident had convinced Jonathan that Peter remained committed to the old format and was unable, for motives of friendship that were perfectly honourable in themselves, to wound people he knew socially.

Honour and decency were qualities that Jonathan valued as highly as would be expected of a man with his liberal education. But – and this is where he differed crucially from Peter – experience had taught him that they were actual liabilities if you tried to deploy them in business. Every city boardroom should be equipped with a metaphorical cloakroom where you could deposit your finer feelings before entering. The trouble was that if you stripped Peter of his finer feelings he would enter the boardroom virtually naked. They were by far his dominant qualities. What he lacked was a streak of ruthlessness, that breathtaking ability to throw loyalty to the wind and turn against people who had regarded him as their ally. TV-am's position, losing hundreds of thousands of pounds a week, was now as desperate as that of a crew shipwrecked on a desert island. The captain had to decide which of them to sacrifice to provide meat for the others to survive. If Peter could not take that decision, Jonathan could. He sharpened his knife and began building the fire under the pot.

He telephoned the institutional directors and invited them to a meeting at his house at cocktail time on Thursday, 10 March. Timothy was there and so were Ben Martin from Barclay's Bank, Charles Wilson on behalf of Jacob Rothschild, Rod Gunner from the Robert Stigwood Organisation and two men representing investors who had come in more recently – Michael Davis from the Prudential and Michael Olsen from Finance For Industry. Jonathan handled them with consummate tact. He showed them Michael's letter. Several times he expressed his high regard for Peter as a human being, qualifying it with his even higher regard for the future of the company and for the millions of pounds that he and the others in the room had at risk. He knew that a move to ditch Peter would create conflicts of loyalty in nearly all of them and, even worse, would provoke publicity that, in the short term, would damage the company commercially. If it were to be done, then it was as well – no, it was vital – that it should be done quickly. There was, Jonathan believed, still time to save the company, but not much. And for those directors not susceptible to the soft approach, beside Jonathan sat Timothy, his Lady Macbeth, the Cassius to his Brutus, ready to drop the veil of decency and put the case against Peter more crudely.

But if Peter did go, who would replace him? Here the Shakespearean precedents still held good. As Brutus succeeded Caesar and Macbeth Duncan, so the spoils would go to the man who wielded the dagger – Jonathan. The conspirators did not see that they had any alternative. Having searched unsuccessfully for a managing director between October and January, they were clearly not going to be able to pick one off the streets in twenty-four hours. It had to be somebody on the board. Dick Marsh, the deputy chairman, was an obvious candidate but the Aitkens had their doubts about him. He would be an abrasive leader in a situation that might call for delicacy. He would be unpopular with the presenters, still rankling over his remark about the Royal Shakespeare Company and 'Listen with Mother'. His administrative experience with a huge nationalised industry – British Rail – and the Newspaper Publishers' Association might not qualify him to fill a position that ideally required experience of television and, because more cash was likely to be required, of raising money in the City. Jonathan had experience, admittedly limited, in both areas. He could be chief executive and Dick chairman.

The substantial argument against Jonathan was that he was an MP and the IBA rules expressly bar anyone in active politics from heading an ITV company. The implications in terms of political bias were apparent. Jonathan believed, however, that once he explained the gravity of the situation to Lord Thomson and his colleagues, and if he insisted that his appointment would be an interim one, for no more than a few months, they would agree to let him do it. The others approved. Ben Martin of Barclay's was given the task of ringing round board members and setting up a meeting for the following day, Friday, 11 March, at Barclay's City headquarters.

Despite the virtual unanimity at that Thursday meeting, at least one of those there remained sceptical that the deed would be done. Timothy left Jonathan's house with Michael Davis of the Prudential. 'They won't do anything, just you wait,' said Timothy. 'They're vacillators. They'll let themselves be talked out of it.'

Peter sensed that events were moving towards a climax. Following the weekend row over Anna and Michael Deakin's confession about his letter, he knew the time had come to make a dramatic assertion of his authority before things flew quite out of control. In deciding the dispute in Anna's favour he recognised, as Michael had done, that he had at last taken sides in the conflict between the presenters and the programme department. In logic he should heed the presenters' increasingly strident demands for better backup from the production crew.

151

And that meant doing something about Deakin. The time had come to bring in a new senior programmes executive to work alongside him, filling the gap Nick Elliott had left when he had withdrawn from the company eighteen months earlier. In discussions with people in the industry two names kept recurring – Mike Townson and Greg Dyke.

Townson, though associated with one of the consortia defeated for the franchise, had never lost his interest in breakfast television. He thought his record in current affairs TV and his experience in popular journalism made him the ideal person to turn Peter's 'newspaper of the small screen' into reality. (Although Peter had talked a lot about the *Daily Mirror* none of his team had ever worked on it and probably seldom read it.) When TV-am were searching for a managing director a few months earlier Townson had put his name forward and had been interviewed by Peter Jay and Jonathan. They thought he would make an excellent director of programmes, but they already had one of those. He lacked the business experience that they felt a managing director should bring with him. Greg Dyke, the up-and-coming editor of London Weekend's 'Six O'Clock Show', had been approached by Michael Deakin about joining TV-am in 1982. He did not react with much enthusiasm when Peter spoke about the possibility again, so Peter made a new approach to Townson. They arranged to meet in the Garrick Club for a drink on Friday, 11 March.

In parallel with that initiative, Peter could see that he also had to do something about the presenters. Although he had declined to dump Anna as unceremoniously as Michael Deakin had urged, he had taken on board Derek Stevenson's view that Michael Parkinson was by far the most marketable of the Famous Five. So he approached Parkinson and asked whether he would host the show at least until the early summer. Parkinson was due to go to Australia in April for six months but he said he would renegotiate his departure date and delay it by several weeks if it was in the best interests of the company.

Derek, who knew nothing of all this or of the meeting at Jonathan's house, was getting desperate about his failure to sell any reasonable quantity of advertising time and was pressing for changes on his own account. On Wednesday, 9 March, he dropped into Hilary's office and asked him how he was planning to make the programmes more attractive. Hilary, caught off guard, could think only of a punchy new series he was putting together about the health service and a hard-hitting investigation of social security fraud. With a sick feeling, Derek could imagine how his advertising clients would react to that. They were looking for something glossy and undemanding. The following evening Derek went to see Deakin and had a long talk about the quality of the programmes and the difficulty of controlling the presenters. 'The trouble with this company,' he observed bitterly, 'is that there are too many friends.'

They were interrupted at around 8 p.m. when the phone rang. It was Ben Martin, calling from Jonathan's house. There was to be a meeting of the board at Barclay's Merchant Bank in the City at 2.30 the following afternoon. The agenda would consist of one item: the position of the chairman and chief executive.

'Have you told Peter?' Derek asked Martin. He had not. 'I think you'd better,' Derek advised, and went to fetch Peter, still working in his office next door. Peter listened to what Martin had to say and scarcely responded. He put down the phone and left, looking grim, without saying a word. He began organising his defences.

Friday, 11 March, 1983, looked like being a typically busy day for Jennie Bland, as she watched Anna and Angela on screen while dealing with breakfast at Abbots Worthy House. She was due on the bench at the Magistrates' Court in the morning, then she had things to do in connection with her work to save a local theatre and her campaign against the proposed route of the M3 motorway. Just before 8 o'clock the phone rang. Sorry to call so early, but the meeting arranged for 2.30 this afternoon has been shifted to 11 this morning. 'What meeting?' Hadn't she heard? There was to be a meeting to discuss Peter Jay's position. People had been phoned the night before. Must have been an oversight. Anyway, there it was, 11 o'clock at Barclay's.

Jennie moved fast. She had never met Peter until she joined the TV-am board but now they were close friends. He would often stay at Abbots Worthy for the weekend, and her children from her first marriage, in the same age group as his, had become friendly also. Until that Friday phone call she had known nothing of the moves being made against him. She had supported him in the October crisis and was gratified when the search for a managing director had been called off in February. She was confident in his ability to head the company. Certainly she would resist any bid to depose him with all the vigour she deployed against motorways and potential theatre demolitions. She cancelled her appearance on the bench, drove to the station and just caught the fast train to Waterloo, arriving in the City in the nick of time.

When she got there she found it was not a full board meeting at all and she was not really supposed to be there. The oversight, it transpired, was not that she had not been informed of the meeting the previous night, but that she *had* been informed that morning. Jonathan had meant it to be just for the institutional directors plus one or two others. They would represent a clear majority of the shareholders but not all of them. Some, in any case, were out of the country. Michael Rosenberg was in Indonesia, where Ben Martin managed to track him down and tell him the news. He was aghast and sent a long telex passionately defending Peter and urging that he should stay on.

Jonathan, still pleading the necessity for speed, was anxious that the whole messy business be concluded rapidly, if possible that day, with an announcement in time for the Sunday papers. But he met unexpected resistance. In two vital cases, men who had attended the Thursday meeting at Jonathan's, representing major shareholders, found that they were at odds with their bosses. Lord Camoys, chairman of Barclay's Merchant Bank, thought his alternate Ben Martin had reacted too hastily in endorsing Jonathan's move. Jacob Rothschild, a contemporary of Peter's at Christ Church, was appalled that his man Charles Wilson should have lent support to a bid to oust his old friend. Rothschild recalled how the Aitkens had originally sought a larger shareholding than the 15 per cent they were allotted, and they had let some of the founder shareholders know that if they ever wanted to sell out the Aitkens would be interested in buying. And people had been surprised by their odd proposal last summer to get a quote for the company on the Unlisted Securities Market.

Rothschild and Jennie spoke powerfully on Peter's behalf. David Frost disclosed the discussion he had with Lord Thomson the previous October, when he was told that the IBA viewed Peter as an integral part of the franchise. Not only would it be wrong to oust him but it might be unacceptable to the Authority. Then Peter was summoned and grilled about his plans to improve the company's dismal performance so far. Even under such a dire threat his powers of persuasion remained impressive. It was bad luck that he could not be precise about all the changes he was planning, but because he was not seeing Townson until that evening it would be premature to mention his name now. Even if he did agree terms with Townson, he saw it as his responsibility to break the hurtful news to Michael Deakin before announcing it to the board. But he did tell them Michael Parkinson had agreed to switch from the weekends and present the weekday show. On administrative matters, Peter said he would institute briefing meetings for non-executive directors every Friday so they would hear the relevant figures for themselves, rather than relying on the press or rumours. He also proposed the appointment of a finance director at board level – the move that he had been urging for months but that had been shelved last October in favour of seeking a managing director.

The directors, who had all now read Deakin's letter to Jonathan, questioned Peter about presenter power, and appeared to find his answers persuasive. Jennie Bland asked him if he had confidence in Deakin. Swallowing hard, and in spite of the notorious letter, Peter said yes, he had. He saw no alternative. Even when he did appoint Townson – or someone else – Peter was not going to dismiss Deakin, just alter his responsibilities. He would still be called director of programmes. In Peter's strict code of honour, it was incumbent on leaders to give public support to their senior aides.

Peter had acquitted himself so well under cross-examination that Jonathan could tell he was not going to be able to rush the institutional directors into an immediate decision, as he had hoped. They agreed to hold a full board meeting on Monday afternoon. Peter believed that by then he would have news about Townson.

Despite the now clear divide between the two factions, the atmosphere remained surprisingly cordial. Afterwards Jonathan felt relaxed enough to tease Peter. 'The trouble with you,' he told him, 'is that you govern too much by consensus. You aren't enough of an autocrat.'

Peter saw a chance of scoring a point. 'Jonathan,' he told him, 'I always try to profit from your advice. But last autumn you wrote me a letter – which I have – saying the trouble with me was that I was an autocrat and I should learn to govern by consensus.' Jonathan laughed gracefully.

A factor that improved the atmosphere, in Peter's view, was Timothy Aitken's absence. An enthusiastic horseman, he was at Tweseldown, near Aldershot in Hampshire, competing in a one-day event. The area was not well equipped with telephones and Timothy had to walk and jog for what seemed miles before he found one. He was like a compulsive gambler frantically trying to discover the fate of the good thing he had staked his shirt on in the 2.30 at Newmarket, with an instinctive feeling that it had been beaten. So when he talked to Jonathan he was not surprised to learn that the coup had failed. He knew Peter possessed devilish powers of persuasion and the board, as he had told Michael Davis, an inexhaustible capacity for dithering. He suspected, too, that Jonathan, with his careful suavity, was a less convincing advocate than he would have been himself. They agreed to discuss their next moves on Monday.

Peter's allies were confident that they had warded off the threat – all except Jennie, one of the many people he spoke to on the phone over the weekend. It was apparent to her, judging from Friday's meeting, that there was no present majority on the board in favour of Peter's removal and that they would therefore decide on Monday that he should stay. But she thought it likely that the Aitkens would try to head off that decision. They would make further attempts to rally support, and would probably need more time to do it. Therefore she saw it as vital that the Monday meeting should go ahead as planned. 'Whatever you do,' she told Peter, 'don't let them postpone it.' So when she was contacted at lunch time on Monday and told that the meeting had indeed been postponed, she sensed that Peter had lost.

Jonathan spent most of that weekend in his Thanet constituency but he had a long meeting with Timothy on Monday to plan strategy. They agreed that one option not open to them was to drop the issue. They had started and they were going to finish. Now they must raise the stakes. Any group of more than 10 per cent of the shareholders in a company, if they feel sufficiently alarmed at the way things are going, have the right to call an Extraordinary General Meeting where voting is on the basis of the number of shares held. The Aitkens had enough shares themselves to summon an EGM but to make the threat credible they needed sufficient pledges of support to give them a fighting chance of winning a majority there for Peter's dismissal. It was a cunning move, for the threat of an EGM is a double-edged weapon. By law at least three weeks' notice has to be given to convene one. During that period, if Peter were to hang on, the company would be paralysed. As a result of the adverse publicity, revenue would decline from a trickle to nothing. The bank would probably withdraw its finance. Even before the EGM was held, the game would be lost. The Aitkens argued that Peter could only be persuaded to quit voluntarily, without an EGM, if it could be demonstrated to him that when it came to a vote, he would lose it. Waverers could be persuaded to the Aitkens' side with the argument that it was the only way to avert the EGM and save the company.

They needed time to put the strategy into effect. They could not afford to risk the enterprise going off at half-cock again. They agreed to have that day's board meeting postponed and to set in motion the technicalities of summoning an EGM. Then between them, armed with this new weapon, they would lobby the institutional investors.

Over the weekend Peter worked to secure his defences. He spoke to Lord Camoys, a key man because Barclay's were not only shareholders but also the bankers who had most at risk. And he went to visit Jacob Rothschild at his home in Warwick Avenue. Peter told him that, after meeting Townson on Friday evening, he believed they could agree terms. That would strengthen what he was now calling the 'remedial package', designed to beat the putsch. Rothschild pledged his support.

The following day Peter called Michael Deakin to his office to inform him of his proposal to bring in Mike Townson. He tried to be paternal and soothing. Like a pregnant mother telling a child not to be put out by the imminent arrival of a baby brother or sister, Peter insisted Michael should not feel threatened. That had no more effect than advising a Christian not to feel threatened when he was thrown to the lions. Not threatened? Deakin knew Townson's reputation as a rough and ruthless operator, whose approach to making programmes was the very antithesis of his own thoughtful, laid-back style. Deakin was furious. His face became deathly pale and distorted with rage. He had heard that some members of the staff were saying it had been treacherous of him to send that 'sneaking' letter to Jonathan about Peter. Treachery? Here was real treachery. Here was Peter going behind his back to appoint someone to share his job. He stormed out of Peter's office and sought out Hilary, to confide in him: then he put in a call to Jonathan.

Thus it was, with the image of Michael's contorted face etched on his mind, and harbouring thoughts of treachery, that Peter drove to Lord North Street for dinner with Jonathan, Lolicia and the countesses.

CHAPTER 6

The Judgement of Jonathan

The morning after the bizarre dinner party, Jonathan and Timothy began in earnest to rally support among the investors for Peter's removal, seeking pledges for votes against him if it did come to an EGM. They had effectively made up their minds to go ahead with this the previous day but Jonathan had seen the late-night walk round Smith Square as a final court of appeal. Had Peter demonstrated a hitherto unsuspected grasp of events, he might even then have convinced Jonathan that he would now take the measures deemed necessary, winning an eleventh-hour reprieve. But the David Frost question remained unresolved. Peter would not give a firm commitment to keep him off the screen in April, when the roster would have him presenting again. And Jonathan did not think Mike Townson, whom he had been told about for the first time that night, was the magic wand Peter seemed to think. Like a hanging judge reaching into his drawer for a black cap, Jonathan passed final sentence. There would be no appeal – though to be fair to Peter, he was much too dignified to consider making one.

Jonathan, Timothy and Michael Scorey, the non-Aitken partner in Aitken Hume, drew up a plan of action. The first step would be to draft the resolution to put before the EGM dismissing Peter, appointing Jonathan as chief executive and Dick as chairman. They would show it to the other institutional shareholders and ask for their support: then they would do the arithmetic. If it seemed that they had a majority for their resolution – or at least if, by virtue of abstentions, there was no majority against it – they would move in for the kill.

They could be certain only of their own votes – 16.7 per cent of the total – and the Prudential, with 5.7 per cent. The Pru's Michael Davis had been a strong supporter of their position, sharing their view that TV-am was poorly run. Among the founders, they could probably count on Michael Deakin and Dick Marsh, with rather more than 5 per cent between them. That gave them just below 28 per cent. At the other end of the scale, the certain 'no' votes were those of Peter and the Famous Five, amounting to 18 per cent. That left a little over half to play for, and more than half of those were controlled by the three largest investors after the Aitkens – Barclay's Merchant Bank, the Rothschild Investment Trust and the Robert Stigwood Organisation, with 9.6 per cent each.

Jonathan knew that Barclay's must be deeply worried about their investment and Peter's stewardship of it. He thought Lord Camoys would not commit himself to a positive vote for the Aitkens' position but might not vote against it, either. Any abstention was useful, for Peter presumably would not want to stay on if it could be demonstrated that he had the support of fewer than half the investors. Following Jacob Rothschild's defence of Peter at the Friday meeting it was unlikely that the Aitkens could count on his votes. The vital ones, therefore, were Stigwood and the three other holders of more than 5 per cent of the equity – Eastern Counties Newspapers, Finance For Industry and Paul Hamlyn's Octopus Publishing Group.

From his discussions, Jonathan had the impression that he could count on Finance For Industry, but the others gave non-committal answers. The most difficult to pin down was Rod Gunner, the young, fresh-faced chief operating officer for the Robert Stigwood Organisation, who represented them on the TV-am board. A keen follower of National Hunt racing and a successful owner, Gunner was at Cheltenham that week for the Gold Cup meeting, the unmissable highlight of the year for the sport's connoisseurs. Jonathan located him on Tuesday, the first day of racing, at the Falcon Hotel in Painswick, not far from the course. He arranged to meet him there for breakfast the following morning.

It was still dark when Jonathan drove quietly away from Lord North Street soon after 5 a.m. that Wednesday. When dawn broke it left a damp mist over the M40, making driving hazardous, but Jonathan reached the hotel in time. Although he enjoyed a hearty breakfast, he could wring no firm commitment from Gunner. The world of show business in which the Stigwood Organisation operates is by tradition emotional and contentious. Gunner's rapid rise in the company was due in part to his very different qualities, his native caution, his preference for keeping his cards close to his chest. He said little at TV-am board meetings but listened carefully to what was going on. He had discerned that the management of the company was unorthodox and he agreed with Jonathan that it needed strengthening. But he was not prepared to make the quantum leap from that position to saying that Peter must go. He temporised. He said he would have to talk to Stigwood, based in Bermuda. He would try to get to the board meeting but he would be staying in Gloucestershire until Thursday, Gold Cup day, the last day of the Cheltenham meeting.

Arriving back in London after a largely fruitless journey, Jonathan had lunch with Mrs Irene Maggs, chairman of his constituency party. He told her he was likely to be busy as chief executive of TV-am for much of the spring, but would not neglect the constituency. In the afternoon he received a phone call from Lord Camoys. 'This can't go on,' Camoys said. 'We must talk.' Jonathan invited him to the Aitken Hume offices in the City. Lord Camoys is from the gentlemanly school of British business. Scion of a landed Catholic family, educated at Eton and Balliol, he would rather see differences settled peaceably than with high drama. He told the Aitkens he was opposed to their move against Peter. He thought there were other ways to fulfil their aim of reinforcing management and strengthening financial control. Timothy and Jonathan disagreed. They said that as the largest shareholders they had every right to call an EGM and that was what they were determined to do. Camoys said that would damage the company and almost certainly provoke the bank into withdrawing credit. Timothy pointed out that he saw no future for the company anyway as long as Peter stayed at its head. Sometimes in business it was necessary to be ruthless. Camoys agreed. He could act ruthlessly if necessary, he assured them. The thought seemed to moderate his resistance. He said that if the Aitkens insisted on going ahead with their scheme, he did not see what he could do to stop them.

They left Jacob Rothschild until last, hoping that by the time they went to see him they would have firm pledges for more than the 50 per cent of votes they needed. They did not, but they thought they could persuade enough 'don't knows' for their cause to succeed. Timothy was deputed to call on Rothschild at his office near St James's Palace. He found the financier unshakeable, having no truck at all with what he saw as a despicable move. But he was Peter's only fully committed ally among the institutional shareholders, and his votes were nowhere near enough. Timothy was confident now that Rothschild's opposition did not matter.

There was one further detail to be attended to. Ironically, they might not have thought of it had David Frost, one of Peter's most passionate supporters, not given them a clue by telling them of his visit to the IBA last October, when Lord Thomson confirmed that he regarded Peter as an important factor in the franchise. Clearly it would be no use forcing Peter's resignation if the IBA would not accept it, so it was important to discover their position. Dick Marsh went to their Knightsbridge headquarters and came away with the impression that they would not interfere. Although they would be sorry to see Peter go, they felt it was an internal matter. Jonathan's position as an MP was an embarrassment but that would be all right so long as he made it clear he intended to remain chief executive only temporarily, for no more than a few months. The last obstacle had been cleared.

On the morning after the dinner party at Jonathan's, Peter drove to the office preoccupied by the realisation that he had to do something to mollify Michael Deakin. They had parted the previous evening on the worst of terms. Even more alarming was the effect their argument had on their mutual friend David Frost.

David's ties with Michael reached back many years. They

had made films together and collaborated on books. But over the last three years David had also acquired a warm regard for Peter, whom he had, after all, brought into the company. They had shared the euphoria of the pre-transmission days and they were mutually supportive in the current worrying times. Now here were his two good friends at each other's throat. He just could not bear it. This was not how things were supposed to be in the Panglossian world he had created for himself in his twenty years in show business, where everything was swell, or, if not, would very soon be.

David's genuine pain made things even worse for Peter and Michael. The row itself had been bad enough, but David's misery was an extra dismal element. Although both were convinced the dispute was over a matter of principle and neither was ready to give way, they accepted David's offer of mediation, mainly to make him feel better. The three met in Peter's office. Peter reiterated that he must appoint a senior man on the programme side and that it was possible the appointee would be of such standing in the industry that he would not agree to work as Michael's subordinate. It did not have to be Townson if Deakin was adamant that they could not work together, but it had to be somebody. Peter proposed that the three of them plus Jonathan Aitken should meet and select someone. He had a dreadful trepidation that he was conceding too much, that the formula he proposed would lead to yet more indecision; but he felt he owed a certain loyalty to David. For a while Michael refused to countenance even that compromise but David persuaded him to agree. They contacted Jonathan, compared diaries and arranged a meeting for Thursday morning.

Came Thursday and Jonathan phoned Peter quite early. He was afraid he could not make the meeting. He had to go to an urgent appointment at Barclay's headquarters in Gracechurch Street. Peter experienced a tumbling sensation from his head to the pit of his stomach. It was evident that the Aitkens were still scheming and he did not see how he could much longer resist. At noon his fears were confirmed. Lord Camoys phoned and asked if he could make himself available at Barclay's at 2.30.

That day he had a lunchtime meeting scheduled with the

board of TV-am News. This includes a number of eminent and worthy people in the news business who gather every few weeks to monitor the progress of the news operation. They had been appointed before the station went on air to fulfil an IBA rule that companies providing news must be formally separated from programme companies. Directors from outside TV-am are Sir Geoffrey Cox, founding editor-in-chief of ITN; Harold Evans, former editor of *The Times*; Andrew Knight, editor of *The Economist*; Sir Gordon Newton, former editor of the *Financial Times* and Charles Wintour, former editor of the *Evening Standard*.

Peter had to leave before the meeting was over. As he went out he passed a note to Andrew Knight: 'I'm going to the City – to be sacked, I think, alas.'

Jonathan decided that it would be appropriate if Lord Camoys were to break the news to Peter. Camoys was unenthusiastic but accepted the argument that neither Jonathan nor Dick Marsh, who were to take over from him, could properly be expected to do it. He took Peter into an office, with Dick in attendance as a silent witness. It was, said Camoys, a painful and difficult task to ask Peter to resign but he had concluded, given the uncompromising attitude of the Aitkens, that it was the only way of saving the company from disaster – not because he, Camoys, thought Peter's regime disastrous but because the factional fighting among directors that would follow the summoning of an EGM would debilitate the company and rob it of its last shred of credibility.

Jonathan had given Camoys a list of the institutional shareholders who would vote for Peter's removal at an EGM. It was, in the jargon of American strategists, a 'best case scenario' including names whose position was far from certain. But Camoys told Peter that even if there were not, in the end, a majority against him, that did not weaken his point about the potential damage the company would sustain from the mere calling of an EGM. Peter wrote the names down in a red exercise book, rather as though he were a reporter making notes for a story on someone else's dismissal. These are the names Camoys gave him, with the amount in percentages of their shareholding:

Aitken Telecommunications (16.7)
Prudential Assurance (5.7)
Merchant Navy Officers Pension Fund (2.6)
Alva Investment Trust (1.4)
Octopus Publishing (5.5)
Robert Stigwood Organisation (9.6)
Finance For Industry (5.1)

That comes to 46.6 per cent. The addition of Dick Marsh, with 3.4 per cent would bring it to 50 per cent. Michael Deakin, if he were to vote against Peter, could ensure victory. Later Peter was to learn that Octopus and Stigwood were a long way from committing themselves. Deducting them from that list would take the Aitkens' positive institutional votes down to 31.5 per cent – but even that might be enough, given that Octopus and Stigwood would probably abstain, as would Barclay's with their 9.6 per cent.

In any case, it seemed to Peter that counting heads was not the crucial factor. Even if only a third of the shareholders were against him, that was a significant minority. He saw no virtue in fighting for his job out of motives that others would construe as personal vanity. He respected Camoys as an honourable man who would not have spoken in such dire terms about the risks of an EGM if he had not sincerely believed them. Peter had been discussing his position over the last few days with his allies on the board. It was Jennie Bland who made the analogy with the biblical story of King Solomon and the dispute between two of his subjects over who was the real mother of a baby. Solomon ruled that the baby should be cut in half and the women share it. One of them agreed but the other said no, she would rather lose the baby than see it killed. Solomon then ruled in favour of the second woman as showing the instincts of a real mother.

It was a crude parable, depending for its validity in this case on how far you accepted the Camoys argument that if Peter held his ground the company would perish. Nevertheless, it gripped Peter. He decided to renounce his claims over the threatened infant. The trouble was that there was no Solomon on hand to mete out the just reward for his act of abnegation. There was only Dick Marsh, sitting there in eerie silence, not at all inclined to assume a judicial role. Peter now realised that he had spent the last few hours preparing himself for what he was about to do, ever since Jonathan had phoned that morning cancelling the meeting about a new programme chief. He had one call to make before reaching an irrevocable decision. It was to John Whitney, the new Director-General of the IBA, asking whether, in the light of what David Frost had been told in October, the Authority would raise any insuperable objection to his leaving. It was the question Dick Marsh had put the previous day and the answer was the same: it was a matter for the company. Peter replaced the receiver and told Lord Camoys he would offer his resignation at a board meeting the next day.

Jonathan had meanwhile been pacing in an adjoining office, like an anxious relative at a hospital desperate to know the result of some delicate surgery. After about an hour he and some lawyers were summoned. They found Peter uncannily calm. Camoys seemed more affected than his victim by the emotion of the moment. The lawyers were there to discuss the all-important severance terms. Camoys had promised Peter he would ensure a generous settlement, and so it was. He was on a salary of £60,000 a year and was offered two years' pay in . lieu of notice, plus the retention of his secretary at TV-am's expense for three months. He also sold back 28,000 of his shares in the company for £5 each – a useful profit seeing that he had paid £1.25 for the original issue and £3.50 for those he bought later. The Aitkens resented these payments and later tried to renegotiate the deal, offering Peter only £25,000 of his compensation immediately, the other £95,000 to come when the company was making a decent profit. Peter accepted that reluctantly but was later able to collect more on a technicality when the Aitkens failed to conform with the small print of the agreement by keeping him informed monthly of the company's results. So he received a further £50,000 in January 1984 with the remaining £45,000 to come in agreed stages later.

While at Barclay's, Peter wrote the short and regretful resignation speech he would deliver to the board the following morning, when he would formally propose that Jonathan succeed him as chief executive and Dick as chairman. Before he left at 6 p.m., word of his demise had already begun to circulate. Both factions suspected the other of initiating the leak. It may have stemmed from the note Peter scribbled to Andrew Knight before leaving for Barclay's. Others may have seen it. Journalists, even those of great seniority, are constitutionally incapable of keeping things to themselves. Howell James, TV-am's press officer, began receiving inquiries from the press while Peter was still at the bank. 'You'd better say you can't find me,' Peter told him. He repeated the instruction when he returned to Camden Lock and went into his office to make telephone calls, one of them to David Frost.

David was appalled at Peter's news and begged him to make a fight of it, even at that late stage. Peter said he could not. He had given his word to Camoys. It had all been arranged. Strictly speaking, nothing was irrevocable until it had gone before the board, but his mind was now settled. David persevered. In two days' time he was due to be married to Lady Carina Fitzalan-Howard, the daughter of the Duke of Norfolk – a triumph of optimism over experience, coming as it did only months after his first brief marriage, to Peter Sellers' widow Lynne Frederick, had ended in divorce. That evening he and Lady Carina had planned a quiet dinner to discuss final details of the wedding. Peter had also been planning an evening *à deux*, a visit to the cinema with a close friend, but David suggested he join him and his fiancée. Peter did not relish the idea of an evening spent listening to David's heartfelt pleading but he thought he owed it to him, as the man who had effectively founded TV-am, the brainchild now in jeopardy from low ratings and boardroom squabbles. They agreed that the four of them should meet at the White Tower, a Greek restaurant north of Soho that had been fashionable in the 1960s, when both men had made their first, sudden impact in the public arena.

At Camden Lock, the building was filling with people. It was the night Terry Farrell, the architect, had chosen to throw a party, chiefly for fellow architects and architectural journalists, to celebrate its completion. It was St Patrick's Night and guests were given glasses of black velvet – champagne and Guinness – before being led on a tour of the building. They were told how the atrium was based on a sunrise theme, symbolising the sun's journey from the east – the Japanese hospitality pavilion – to the deserts of the American west at the other end, encompassing the Mesopotamian staircase on its journey. They marvelled at the wealth of potted plants and trees and at the main studio, with its battery of overhead lights that some TV professionals thought overelaborate. Among the guests was Richard Francis, managing director of BBC Radio, basking in praise for 'Breakfast Time', telling admirers how it had all been done on a shoestring at Lime Grove. This extravagance – gesturing towards the greenery – was all very well, but. . . .

It was not long before the guests had something meatier to gossip about than the architecture. Just after 7 p.m., the Channel 4 news programme, compiled by ITN, had reported a strong rumour that Peter was about to resign. All rival bulletins are routinely monitored by any news-gathering organisation and within minutes the rumour began to sweep the party. Then Peter appeared at the top of the stairs, looking uneasy and carrying a briefcase. He was leaving for his dinner date but to reach the door he had to descend the steps and force his way through a milling crowd of drinkers. He had been invited to the party, so before descending he surveyed the mob to locate the host, Terry Farrell, then pushed his way towards him.

'I hope you'll forgive me if I don't stay,' he said. 'I've had rather a bad day.'

At 8 o'clock, the Channel 4 news bulletin ended with a confirmation of what it had reported as a rumour earlier. Michael Deakin called as many of the staff as he could locate for a meeting in his office. He announced that the rumours were true. 'After all,' he explained, in his precise, high-pitched voice, 'the shareholders are the bat, ball and stumps in this game.'

With the news now certain, the atmosphere at the party changed. Jubilation is not the right word, but a certain sense of satisfaction, even triumph, rippled through the guests, especially those in the TV business. Peter has often been accused of being arrogant and he resists the charge, but it is an impression he inescapably gives. And it was not just Peter. The whole approach of TV-am, the glitter and the self-confidence, the implied scorn for existing news values and techniques, had grated in an industry where, of the seven deadly sins, envy may be the most potent. Sympathy for Peter and the Famous Five – and there was plenty of that – was diluted by the feeling that they had it coming. The party was due to end at 8.30 but at 10 o'clock a dozen or so guests remained, reluctant to tear themselves away from the scene of such momentous events. They drifted upstairs into Michael Deakin's office to watch the ITN news and hear the facts confirmed yet again.

The dinner for four at the White Tower was an emotional affair. They largely ignored the food and the stares of fellow diners as David argued over and over again with Peter about whether he should make a fight of it. David pleaded with Peter to withdraw his agreement to resign and pledged his full support, even his own resignation if necessary. Peter was deeply touched and said so. But he reiterated that he could not go back on his word to Camoys or stay on without the full support of the major institutional shareholders; and if he did stand up and fight, forcing the Aitkens to summon an EGM, the row could well destroy the company. Peter had been much taken with Jennie Bland's story about the judgement of Solomon and kept repeating it. He and David were handing the imaginary baby back and forth over the tablecloth. When it came to coffee time and Peter had still not been persuaded to reassert his parentage, David at last began to sense defeat. But nonetheless he resolved that he would, at the next day's board meeting, make a gesture of support for Peter. If he were seen to accept the situation unquestioningly, he would never be able to forgive himself.

David was the only one of the presenters Peter had confided in. The others had to rely on the news reports and rumours. When Anna and Angela arrived at the studio, therefore, at 4.30 on the morning of Friday, 18 March, they knew that Peter was going to be asked to resign that day and was expected to do so. They did not know that he had agreed to go and would not countenance any attempt to make him change his mind. Had they known that, it is unlikely that they would have embarked on the public and dramatic course they now took.

As they climbed from their cars in the early morning darkness, they were greeted by a cluster of reporters, well wrapped against the cold, and by a crew from BBC Television News. Angela arrived first. She can be reticent with the press, but on this occasion was glad to talk.

'I'll tell you, Peter Jay is being made a scapegoat. He's not the man who should be going.'

'Do you know that there's been an attempt to make him go?'

'Yes we do. We've known it for some time in the building.'

'So far as today's board meeting is concerned, will you and your colleagues, David Frost and everybody, be doing anything about it?'

'Well, we're not all on the board, of course. David Frost is on the board. He knows how we all feel and I hope that there are enough people who work on the board who know enough about television, the presentation of television programmes, who will actually support our point of view.'

When Anna arrived a few minutes later, she was equally ready to give forth.

'I think that Peter Jay has the full confidence of the five major presenters and I can say that with some confidence myself.'

'Your colleague Angela Rippon has said here this morning that there's somebody else who should be going rather than Mr Jay.'

'Well, I think there's been a great deal of treachery in the company and unfortunately because of the bounds of our contracts we're not allowed to talk about it. We would love to. I think history will expose those who have been most treacherous and I think that it's come at an extremely unfortunate time. This company will flourish, I'm sure of that, and I think that's what we should be saying and what the board should be saying.'

The two presenters were keen to know the precise situation. Now that they had made their protest public the question arose whether they should say anything about it on the programme. Unable to reach Peter, Angela phoned Howell James, the press officer, at home. But he had been told nothing. She and Anna went round the building trying to drum up support for Peter. They found an enthusiastic response. Despite his distant manner and his unpopularity with some others in television,

Peter did inspire loyalty among those who worked for him. Anna opened the show: 'Good Morning, Britain. It's Friday the 18th of March. We're all in a very good mood here this morning, with decision day in the Darlington by-election less than a week away and other decisions going on here behind our backs.' She said the last nine words hurriedly, as though anxious to get them out of the way, like a cheeky schoolgirl being rude to the head in assembly and knowing she will be punished for it.

Angela waited to make her contribution to the cause. She was talking about the morning papers to Nicholas Scott, a junior minister (the man who nine years earlier had been named with Peter as one of *Time* magazine's future world leaders). 'There's one story in the newspapers this morning we're all interested in,' she said. 'You know we have a soap opera here at TV-am which is called "The World of Melanie Parker". I think we've all decided actually that our own soap opera would out-soap "Dallas" but that's another matter altogether.' During the interview with Scott, Angela made the bizarre suggestion that she and Anna might get involved in mud wrestling in the canal, presumably as an alternative form of employment if the worst came to the worst. Nick Owen, the sports presenter, put in his two-ha'porth by referring to 'Camden Lock's own fun ship'.

Anna moved in deliberate stages towards her major *démarche*. They reached the spot in the programme where the pair were reading letters from viewers. Anna said somebody had written in recommending that instead of reporting what the papers said they should reveal what the papers *don't* say. She broke with custom by not revealing the name of the originator of this especially timely notion, which allowed her to comment: 'I think it's an incredibly good suggestion because it is in fact what the papers don't say which is the most interesting thing. There's an awful lot the papers aren't saying about TV-am at the moment. We'd love to tell you but I'm afraid we're honour bound not to. Perhaps later we shall reveal all.'

Then Angela hit on the idea of looking at their horoscopes. Both are Librans. 'All forms of communication are highlighted,' she read, 'and there will be much to interest you in today's happenings – crikey, I'll say. An evening meeting will put useful information your way – I shouldn't have started this, should I – perhaps a secret tip and an unexpected weekend invitation could give a fillip to your mood.'

A bit later Anna listed what was to come up in the second half of the programme. 'We'll be talking to singer Sandie Shaw about the world of pop music and Zen Buddhism and Nick Scott will be telling me what he found in this morning's papers. One of the stories in this morning's papers is of course about intrigue at TV-am. High dramas going on in the boardroom this morning, and we gather apparently that our chairman Peter Jay is going to be asked to resign. We want to send him a message from everybody here this morning simply saying: "Peter, if you're watching, don't resign. We like you very much. We'd much rather you stayed!" '

After the programme Anna and Angela would normally go quietly to their offices but today they walked into the centre of the ground-floor administrative area and stood at the foot of the Mesopotamian staircase. Clive Jones, the managing editor, tried to moderate the growing sense of drama by inviting journalists to a meeting in his office. He was anxious they should not be swayed by the passions of the moment into taking a stand they would later regret. But that still left enough staff to follow the two presenters into the courtyard and pose for a thumbs-up press picture to show support for Peter. Some carried hastily-scrawled bills reading JAY MUST STAY. Anna and Angela spoke in much the same terms they had used on the way in earlier, although now Anna had refined the script and was talking of 'acts of enormous treachery'.

Her on-screen appeal to Peter to stand fast did not reach him and would have had no effect if it had. By the time she was saying it he was driving to an 8.15 meeting with his successors Jonathan and Dick at his lawyer's office in Piccadilly, to tie up loose ends about his severance terms and his company Volvo. Jonathan was surprised by Peter's sang-froid, his persistent cordiality. Although the victim vouchsafed that he had not eaten the traditional hearty breakfast, he had enjoyed several hours of untroubled sleep. They drove separately to Gracechurch Street.

. Michael Deakin and Derek Stevenson left Hawley Crescent together for the board meeting. They were jostled by the crowd of reporters outside who asked Michael: 'Are you the traitor?' He looked pained and said nothing. Lightheaded, Derek jauntily whistled the Marseillaise. He was still whistling away when they reached the bank, until Jacob Rothschild asked him to desist.

Jennie Bland had been phoned about the meeting at 10 o'clock the previous evening. Apart from the principals in the drama she was the first to arrive and sat by herself in the boardroom for a few moments while Peter and Jonathan conferred with Lord Camoys in another room. Camoys entered and led her to a third room where he told her: 'I hope you're not going to be difficult about this.' He explained that the bank would withdraw its financing if Jonathan carried out his threat to summon an EGM – not his bank, Barclay's Merchant, but the main Barclay's Bank which was underwriting much of the loan facility. He too was upset at the way things were being done but Peter had accepted it and it would now be best for everyone if the nasty business coud be concluded as speedily as possible.

Jennie went back to the still empty board room where she was joined by Peter, seeking to reinforce Camoys' plea. 'Look Jennie,' he said, 'I want to ask you to go along with this. I've accepted it. I've said I'll go. I won't go back on it. If a large portion of the board think I'm not doing a good job it's right that I should go.' He was worried that she was going to reintroduce that judgement of Solomon theme and urged her not to. 'It will make it difficult for me if you do. Let's go through it without a lot of aggro.'

Despite these powerful pleas, Jennie was stubborn. 'I feel too strongly about it,' she said. 'I'm not just thinking of you. It will be bad for the company if you go. Someone will take it over who doesn't care about it the way you do. It will mean bad publicity and an inferior new management. We should be putting right what's wrong instead. I don't think I can do what you ask.'

Peter, angrier with Jennie now than with Jonathan, sat next to her as the directors drifted in for the meeting. He was not even to be allowed to resign gracefully. There would be a scene. Yet even if Jennie had agreed to keep quiet for his sake, there were two others who would not – David Frost and Jacob Rothschild. Though David was sure, after the White Tower dinner, that there was no chance of persuading Peter to change his mind, he felt it would be wrong not to place on record his distress at what was going on. Michael Parkinson had gone to Barclay's and wanted to address the board but was not allowed to; so David, realising he was representing all five presenters, delivered a speech steeped in emotion. He praised Peter without restraint and defended his record at TV-am. It was quite unfair, he said, to judge his performance after only six weeks of broadcasting. They had always known things could go badly wrong at the beginning. Peter, who had devoted two years of his life to selfless service of the company, was the man most equipped to put them right. The board were being rushed into a decision, and it was a wrong decision.

Jennie, moved by this appeal, felt she could not let David carry the burden alone, so she delivered a speech in similar terms. But it was Jacob Rothschild who made most impact with the directors. Peter, in the chair, had repeated in response to Jennie and David's speeches that their arguments were insignificant when set beside the colossal damage the company would suffer if an EGM were called. Rothschild said the way to avoid that was not for Peter to resign but for Jonathan to withdraw the EGM threat. Jonathan, with Timothy sitting watchfully at his shoulder, said he could not do that. From time to time Jonathan would brandish his EGM resolution and make as if to lay it before the board. 'Put it away, put it away,' cried an appalled Lord Camoys.

Ironically, had it not been for Peter's determination to quit, the board would probably have voted for him to stay on, so powerful had Rothschild's advocacy been. Gordon Cartwright, representing Paul Hamlyn's Octopus Group, said he had been instructed to make it clear that Hamlyn did not support the Aitken move. Michael Rosenberg, still abroad, sent another long telex in Peter's defence. Michael Deakin was one of the few who spoke for the Aitken line. He told the board about Anna's and Angela's pro-Jay demonstration that morning, citing it as another case of the cancerous 'presenter power'.

The meeting lasted until nearly 1 p.m. Given Peter's dignified insistence that he would quit and his refusal to fight a rearguard action, the directors had no realistic alternative to accepting his resignation. At Peter's suggestion there was no vote but the resolution was passed *nem. con.*, meaning that although not all the directors were positively in favour, none opposed it sufficiently to vote against. Rothschild went along with that only on condition that his objections were recorded in the minutes.

It was over. Peter slipped out of a back entrance and took his solicitor to lunch at the Garrick. Jonathan and Dick Marsh walked through a tunnel linking Barclay's Merchant Bank with the headquarters of Barclay's Bank proper, where reporters had been waiting for the best part of the morning to hear the outcome of the meeting. Dick read out Peter's prepared statement and would not add to it. Jonathan said his appointment as chief executive would be temporary, pointing out that as an MP he would not be allowed by the IBA to hold it for long. He clearly could not announce any immediate rescue plan until he had had a chance to study the company's position in detail, but one of his priorities would be to look at the presenters' contracts. Asked about Anna's and Angela's fiery statements, Jonathan attributed them to the 'supercharged emotional atmosphere'.

David Frost, still upset, issued his own statement after the meeting giving vent to his total opposition to Peter's removal and adding: 'I would like to place on record that I think his contribution was magnificent.' Back at Camden Lock, the TV-am branch of the National Union of Journalists passed a motion of confidence in Peter.

Jonathan took Dick to his Lord North Street house for lunch. At 3.30 they drove to Camden Lock, besieged by reporters. The first request to the new diarchy was for a meeting with the presenters – all except Angela, who had already caught her train to Devon for the weekend. Jonathan invited them to Peter's old office, now his, where he sat incongruously in front of the wall containing the photographs, trophies and mementoes of Peter's brilliant career. The meeting began bad-temperedly, with Anna asking him brusquely what qualifications he thought he had to run a TV company. The presenters, except David, angrily blamed Michael Deakin for the poor programmes. Jonathan was soothing and soon managed to lower the temperature. He urged them all to stop making mischief by talking to the press. Instead they should work calmly and normally while he devised a way of rescuing the hard-pressed company, where a lot of livelihoods were at stake apart from theirs. (His appeal for lower voices was not heeded by his cousin Timothy, who had lost no time in telling reporters that the crisis had been provoked by the presenters' usurping control of the company. 'It was very much a case of the tail wagging the dog,' he said.)

When the meeting was over, Jonathan reflected that he had been given a smoother ride than he expected. Nobody had threatened to resign – though he felt it might have eased some of his problems if they had.

While the meeting was in progress Peter arrived back in the building, wishing to explain matters to his secretary. At the top of the stairs he looked through the glass walls into his office and saw Jonathan in what, until a few hours earlier, had been his chair, with the others milling about. For Peter that was possibly the single most painful moment of the entire drama. In the last few days he had been preoccupied with the politicking. He had no time to reflect on what his departure would mean in terms of his day-to-day routine. Now here he was, face to face with the prospect, seeing his office occupied by someone else and the company's business being conducted without him. He turned on his heel and walked away. By a cruel coincidence, his new company Volvo had been delivered the previous day. He climbed into it and drove to Oxford, where his eleven-year-old son Patrick was appearing in a school production of *The Wizard of Oz*. Peter refused to speak to the press at the weekend but broke his silence on Monday in an interview with the *Standard*, where he displayed that ability to detach himself from events that the American novelist Nora Ephron had detected in him. 'I never feel bitter, ' he told the reporter. 'It is a destructive and unworthy state of mind.' He added he was sure he would remain a personal friend of Jonathan.

With the presenters defused for a while, Jonathan and Dick turned their attention to the rest of the staff. They spoke to the executives and the union representatives, then called the remainder together and urged that old enmities be buried and everyone get down to improving the programmes. Jonathan said there would be changes but before he made any he would seek the views of the staff and would welcome their ideas. He might ask sacrifices from the unions. He was determined the company would survive.

That evening he stayed late at Camden Lock and gave a few off-the-record press briefings, the fullest to Richard Brooks and Marjorie Wallace of the *Sunday Times*. He was now prepared to go into greater detail than at his press conference about what he saw as Peter's failings. Relaxing languidly among his predecessor's trophies, he described the budget crisis of the previous autumn and told of what he saw as inadequate financial control. He criticised Peter for indecision and for failing, in his view, to take charge of the presenters, creating virtual anarchy. TV-am was 'a sinking and unhappy ship'. He would bring in younger presenters, stabilise the programme by making fewer changes to the roster and inject a more popular approach.

The interview was interrupted at 10 p.m. because Jonathan wanted to see how ITN were reporting the news about TV-am. There was a moment of near-farce as he pushed the buttons of the unfamiliar set ineffectively and had to seek the help of his visitors to switch on. A TV executive unable to operate a TV set could, he reflected, be the butt of satire. He found the news about TV-am was still being given prominence on the bulletin, with pictures of Anna and Angela rallying support for Peter outside the building, and of David speaking up for his friend. Jonathan could see that the constant repetition of those emotional scenes would damage the company's commercial credibility, adding yet another problem to the dozens that, after only a few hours in his new post, he realised he was facing.

The sense of action-packed melodrama was heightened next day by David's wedding. A chap cannot be blamed for getting married but he could not have chosen a less apposite time. Although no formal announcement had been made, word had been circulating for days that he and Lady Carina would be at Chelsea Register Office on Saturday and so they were, with the bride in a traditional lacy wedding dress carrying a bouquet

177

and wearing on her blonde head a floral circlet. Michael Deakin was there and so were scores of reporters and photographers, holding up traffic in the King's Road. Nobody would say anything about TV-am as they left the Register Office for the reception at Les Ambassadeurs in Mayfair, where Michael Parkinson was among the guests. Another party, another affirmation that, despite the boardroom ructions, life went on. As Parkinson had said on the air that morning: 'It's business as usual under the new management.'

In the first few days in his new job Jonathan received many telephone calls of congratulation – or, in many cases, thinly disguised commiseration. One came from New York, from Richard Nixon. The two had been friends since the early sixties, when Jonathan had worked for the British Foreign Secretary, Selwyn Lloyd, and Nixon was a defeated presidential candidate. In 1979 Jonathan and Lolicia had spent their honeymoon at Nixon's house at San Clemente, California. Now Nixon phoned to say he had just read about Jonathan's new appointment. Congratulations and what was it like?

'You know that book of yours, *Six Crises*,' Jonathan replied. 'Well, it's like that, but all six crises are happening at the same time.'

The most serious of them had developed suddenly over the weekend. It involved the IBA. Although John Whitney had given the go-ahead for Jonathan to take over as far as IBA staff were concerned, the final decision lay with the twelve Authority members. It is possible that they would have accepted Whitney's recommendation without demur had it not been for some weekend agitation by a few MPs and a trenchant leading article in the *Observer*. The Labour left-winger Frank Allaun was the first to protest. He had clashed with Jonathan in the Commons only a few weeks earlier, when Allaun's private member's bill to give people a right to reply to press criticism was defeated, partly because of Jonathan's outspoken opposition to it. Now, on the day of Peter's resignation, Allaun declared: 'The IBA should intervene. Mr Aitken is a Conservative MP who is going to exert influence over radio and TV. How can he do this objectively?' Allaun said he would raise the matter with the Home

Secretary, Mr William Whitelaw. Jonathan replied: 'I want to assure everybody that my politics will not interfere in any way with the editorial content of TV-am. I have been a journalist for a long time and I value freedom of speech too much to allow my own politics to interfere with it.'

That Sunday the *Observer*'s main editorial was headed THE IBA'S DOG'S BREAKFAST. It began:

> The IBA has never been a particularly satisfactory body. Seldom, however, can it have displayed itself to greater disadvantage than in its reaction to the boardroom dénouement at TV-am.

Tracing the history of the franchise, the *Observer* found 'singularly little evidence that it [the IBA] had any real understanding of all that was involved' in breakfast television. They had awarded the contract to an 'essentially showbiz consortium' with the apparent objective of brightening up their own dowdy image.

> If the IBA started off by being casual as to its responsibilities, it has ended up by being cavalier as to its duties. As a statutory regulatory body it simply is not enough for the Authority to state that the machinations and convulsions in the TV-am boardroom are 'primarily a matter for the company's board to determine' [a quotation from the IBA's Friday statement]. That would be an inadequate response in any circumstances: where the company involved has been brought under the control of a Conservative Member of Parliament and a prominent Labour defector who is today a declared Tory, it represents nothing more than a supine abdication of the role for which the IBA was established by Parliament. ... Mr Jonathan Aitken's position as TV-am's new 'acting' chief executive – while remaining a backbench Conservative MP – is, in particular, wholly indefensible.

It was, the editorial continued, 'actively dismaying' that Whitney and Lord Thomson had failed to recognise this.

> The main hope must now be that the dozen lay members of the Authority will yet demonstrate a rather sharper sensitivity not just to a question of protocol but to an issue of principle.

Thomson, who returned at the weekend from an overseas visit, summoned a meeting of the IBA for Wednesday, 23 March. Jonathan received a call from the IBA saying that it

179

looked as though the members, rattled by the stern *Observer* editorial, would not ratify his appointment. He would have to stand down right away. His first reaction was to accept that verdict but Timothy, who was with him when the call came through, was appalled and angry. He persuaded Jonathan to phone Lord Thomson right away and demand the right to be heard. Thomson was already at the board meeting. Jonathan told Thomson's secretary that he was coming straight to Knightsbridge to talk to members of the Authority. He hoped no irrevocable decision would be taken until he arrived. Timothy went with him, but Jonathan did most of the talking. He said a further change of leadership at TV-am so soon after the recent upheaval would have a damaging effect on staff confidence and could bring the company to the point of collapse. The Authority were suggesting that until a chief executive was found the company should be run by an executive committee of the board. That would be disastrous, Jonathan told them. Any improvement in morale in the last five days was due to his being there, making decisions and being seen to make them. An executive committee would never be able to decide anything. It would be an abandonment of responsibility. He would prefer to have Peter Jay back. Jonathan agreed that in principle it was unsuitable for an MP to head a TV company but this was an emergency. He would give them his personal guarantee that no political bias would creep into the programmes. And he would stand down just as soon as a suitable successor was found.

With some reluctance, the Authority succumbed to his pleading. He could stay for a while, but for weeks rather than months. They set a deadline of 11 April – the end of Parliament's Easter recess. That gave him less than three weeks to make the changes he deemed essential and to find a new chief executive. The Authority made a further proviso. Bearing in mind his limited tenure, they asked Jonathan not to take any controversial and potentially damaging actions, especially no dramatic dismissals. Further blood-letting, they felt, would lead to more deplorable publicity.

Among the first fruits of Jonathan's plea for suggestions from the staff was a long memo from Derek Stevenson detailing what, in his view, was wrong with TV-am. It read like a catalogue of the sort of grouses against life in general to be heard in the saloon bars of pubs, but it did give Jonathan something on which to base his intended reforms. Stevenson complained about the debilitating effect on the company's finances of the Camden Lock building, the lack of experienced management and inadequate financial controls. The shift system was unbalanced, with too many people working at night and too few during the day. There was no philosophical document to describe the programmes that should be made. 'Many people believed the presenters would see us through whatever we put out on air.' And he repeated one of his favourite criticisms: 'Too many friends'.

There is no doubt that the problems Jonathan found when he arrived were substantial and grave, yet in assessing them it is necessary to take account of the 'look-at-the-mess-our-predecessors-left' syndrome. Jonathan is, after all, a politician. When new governments take office they invariably affect horror at the deplorable condition in which their departed opponents have left the nation's affairs. They stress the failures, sniff out and publicise cases of apparent mismanagement and alleged extravagance. Any successes the old regime may have achieved are either ignored or claimed as the work of the new. Jonathan and Timothy set about criticising Peter and his works with enthusiasm, their case strengthened by the unarguably abysmal audience figures and financial performance.

An easy charge to make – and among the hardest to refute – is extravagance. In most organisations you do not have to search far to find an office car more luxurious than it need be (Michael Deakin's Porsche), travel expenses inadequately justified, expensive-looking pot plants cluttering the work space. Dick Marsh was delighted when, searching one of Peter's cupboards, he found it stuffed with bottles of champagne. There! Living it up while the ship sank. Dancing aboard the Titanic. The story was retailed with malicious glee until David offered an explanation. On the opening morning in February many wine merchants had sent crates of free champagne for the launching. Not all of it was used and Peter, frugal in many ways, locked it in his office in case there was ever again anything to celebrate.

One of Jonathan's first decisions was to ask Timothy and Roger Frye, Aitken Hume's finance director, to make an accurate survey of revenue and expenditure. They found things worse than they had suspected. Expenditure, they calculated, was not far below £2 million a month, with ongoing revenue scarcely more than £300,000. With that kind of deficit, bankruptcy could not be further than weeks away. The figures provoked panic at Barclay's. When Lord Camoys was persuading Peter to stand down, one of his most forceful arguments was that if the boardroom struggle were allowed to develop into an EGM, the banks would withdraw their finance. Peter resigned to forestall that outcome, but a week later the banks were in any event reconsidering their position in light of the alarming figures. Barclay's Merchant Bank had hitherto been offering overdraft facilities of £3 million and Barclay's clearing bank a further £1.45 million. The Merchant Bank now reduced its facility to £2.25 million and the clearing bank withdrew its commitment altogether.

Jonathan now had the problem of maintaining the confidence of the other investors, for he could see he would shortly have to ask them for a lot more money. His difficulty was compounded by his and Timothy's frankly combative style of making what they liked to call 'total disclosure' of the financial position they inherited. They argued that if the unions were going to be persuaded to make necessary sacrifices they would have to be presented with the full desperate picture. The snag with that practice was that the investors also got to hear of it and grew alarmed. It became necessary to initiate convincing cost-cutting moves while trying to do something about the revenue. Higher revenue could only come from higher ratings, resulting from better programmes. That meant that the programmes must be his priority. The presenters insisted the failure was the fault of Michael Deakin, Hilary Lawson and the programme staff, while Michael and Hilary blamed the presenters. Jonathan thought both views contained part of the truth. Michael, old friend or no, despite his support for the Aitken coup, had to be swept aside, along with his protégé Hilary. A new programme editor must be brought in and placed out of Michael's range. Jonathan's first notion had the merit of simplicity. He asked David whether he would take over as editor. This would solve two problems at once, the second being how to get David off the screen. David thought about it seriously but, flattered though he was, decided against it. To agree would be to concede that his regular on-screen appearances would end. That was the part of the business he loved and he knew he was best at, despite the recent adverse criticisms of him.

With David out of the running, Jonathan received a call from Alastair Burnet, the ITN news reader, proposing Mike Townson, the man Peter had been flirting with just before his departure. Jonathan asked others in the business about him. 'Townson? A bit of a thug,' said one, referring to his reputation as a tough though highly professional operator, not to any personal characteristics.

'Just what we need,' Jonathan replied. He told friends it was time to leaven Michael's Oxbridge intellectual elite with, figuratively speaking, some sweaty hacks. Jonathan interviewed Townson but found he wanted to be a managing director rather than the editor TV-am needed.

The obvious choice would have been the BBC's Ron Neil, in the hope that he could work the same magic with 'Good Morning Britain' as he had with 'Breakfast Time'. But like most BBC people, he proved impossible to dislodge from the benign Corporation. In the end Jonathan plumped for Greg Dyke, younger than Townson, a bit less 'sweaty' and without his boardroom ambitions. Dyke had earlier been considered by both Peter and Michael but had fought shy. This time, given assurances about his autonomy, he swallowed the bait, despite attempts by London Weekend to persuade him to stay with them. His contract was a challenging one. He would be paid £40,000 a year to begin with. If the ratings later climbed through the million barrier, his salary would go up to £60,000. Although technically Greg, as editor-in-chief, was junior to Michael, the director of programmes, it was agreed that Michael would not exercise authority over him. Michael was removed from day-to-day programming and made responsible for the children's output, video sales and some of the diplomatic functions Peter used to perform, especially relations with the IBA. Hilary was moved from the programme department to become Jonathan's deputy managing director – a post he held for only a few weeks before resigning from the company.

Dyke was not going to arrive until May, after working out his notice with London Weekend. Jonathan was anxious that he should find a reasonably clean house when he came and that meant revising the expensive and inefficient shift system. Working closely with managing editor Clive Jones he devised a plan to cut the former four shifts to two and substantially reduce the number of people working at night, covering non-happening news. Instead, more would come in during the day to work on features. There would be less confusion and less overtime worked.

A more sensitive area of house-cleaning concerned the presenters. Dyke had said in his initial interviews with Jonathan that he would prefer to work with a fresh slate of people. As Jonathan saw it, leaving aside whether the Famous Five were a liability from the ratings point of view, they were certainly a financial drain on the company. He reckoned that, if you took into account their fringe benefits and appurtenances − David's gag writer, for instance, and Michael Parkinson's wife, who helped him host the weekend shows − the five of them cost £600,000 a year. He was sure it was more than they were worth.

On 28 March Jonathan spoke to four of the five individually (David was in New York), some at his house in Lord North Street, and asked them to take salary cuts. The men agreed to reduce their take by about a quarter but the women would not. Anna said £75,000 was her market value and she would not negotiate it downwards. Angela, who knew she was lower paid than the others, said she would only think about taking a cut when the rest came down to her level. Since Michael Parkinson and David Frost were being paid nearly double her £60,000, that was an unrealistic condition.

Shelving the question of pay cuts, Jonathan sounded out the presenters on changing their duties. He insisted that David, due to go back into the roster at the beginning of April, should stand down, concentrating instead on 'the big interview'. David, still in New York, had no choice but to accept. It was embarrassing for him but he put on his usual bold face. A week earlier, before flying to New York on Concorde with his new wife, he had told reporters: 'There is to be no axing from TV-am. No one is going to be kicked out. At the beginning of April I will be back on the programme.'

When he returned from America the position was not as he had stated. In the intervening period it had been announced that he was being replaced as a presenter by Nick Owen, the former sportscaster, whom the press, still obsessed by the story, hailed gleefully as 'Mr Ordinary'. But David continued to look on the bright side.

'There is no question of being disappointed,' he said at Heathrow Airport. 'The thought – the joyous thought – of not getting up at 4 a.m. is the most delicious wedding present I could have.'

Now Jonathan had to tackle the women. In the immediate aftermath of the coup Anna had said she would do almost anything to help the station. 'I'm sure I would be prepared to do a newscasting job. I'm also quite good at cleaning and polishing. I don't expect to be sent to the salt mines.' Jonathan was not suggesting any of those but he put it to her and Angela that they might get out on the road as reporters, doing special shows from the regions, in-depth reports, that sort of thing. Neither was prepared to commit herself immediately. Michael Parkinson, clearly the most successful presenter, was not involved in a job switch; while Robert Kee, who had appeared less than the others so far, was about to take a long break.

For the first week in April, Angela and Nick Owen hosted the show. Ratings remained in the doldrums and now the few remaining viewers had something fresh to complain about. According to the *News of the World*, they believed Angela was being nasty and patronising to Nick. 'Mr Ordinary' was becoming 'Mr Henpecked'.

'We've been inundated with calls,' a member of the TV-am staff was quoted as saying. 'People think she's trying to put Nick down all the time, treating him sarcastically.'

She was said to have told him once: 'There you are, it isn't as easy as you think, is it?' On another occasion, when Nick gave an impersonation of the station's cartoon character, Roland Rat, she suggested that the two swap places, with Roland presenting the programme. A friend of Nick's said he was 'in despair' and a colleague of Angela's was quoted as calling her 'a bit of a monster'.

Whether or not that was the reason, in mid-April Angela was replaced by Lynda Berry, who had been reading the news on Daybreak. When Angela heard about it, returning from a week in Australia, she was angry.

'I don't think this is the right moment to get involved in a slanging match,' she said. 'There will come a time, though – and it may be quite soon – when I propose to have my say.'

As Angela seemed to be predicting, her relations with TV-am were rapidly nearing a showdown, but it was not one that Jonathan could stage. His bargaining position with the presenters was fatally weakened by the IBA's edict barring him from firing them. It was something he would have to leave to his successor; yet he was making no progress in finding one. The executive head-hunters had produced no more thrilling candidates than they had the previous winter. Almost the only interesting outside application came from Sir Larry Lamb, the former editor of the *Sun*, now back from a spell in Australia and job hunting. (He would later be appointed editor of the *Daily Express*.) Sir Larry proposed himself to Dick Marsh but Jonathan thought it would be a mistake to appoint another journalist. Although Sir Larry and Peter possess very different qualities – as both would be eager to point out – Jonathan thought that at this precarious stage in the company's history the requirement was for a rough businessman who knew his way round the City institutions that would certainly have to be approached for more money before long. For the same reason he rejected another suggestion, to offer the post to Charles Wintour.

Timothy and Michael Scorey tried on several occasions to persuade Jonathan to sacrifice his seat in Parliament and stay on at TV-am. There was a precedent. Aidan Crawley had done that when London Weekend, of which he was chairman, won its franchise in 1967. For a time Jonathan considered it seriously. He saw little prospect of political advancement under the present leadership, despite (or perhaps because of) the close friendship he had enjoyed with Mrs Thatcher's daughter Carol before his marriage. Yet Mrs Thatcher would not last for ever and a Parliamentary seat was not something easily gained or to be given up lightly. Timothy and Scorey did not seem to appreciate how much it meant to him. Besides, he had given his word to his constituency party chairman that his fling with TV-am would lead to no permanent liaison. He did not want to go back on that. So he said no, then neatly turned the tables. Why didn't *Timothy* take over? He had been involved with the company since its formation, even if spasmodically. As a result of his inquiries into the financial position he knew where any bones were buried. Like Jonathan, he had the media in his blood; and he was coloured with that streak of ruthlessness that both were sure was needed if the company was to be led to prosperity.

At first Timothy was horrified at the suggestion. As chief executive of Aitken Hume, charged with establishing it among the City's leading finance houses, he had plenty on his plate. He liked what he was doing and thought he was doing it well. Why should he swap control of a burgeoning young company for one that was close to collapse?

But Jonathan was persuasive. He pointed out that the Aitkens were now publicly committed to the success of TV-am. If it were to fail, part of the blame would, however unjustly, fall on them, and might affect the financial community's attitude to their other enterprises. By engineering Peter's downfall they had assumed at least a moral responsibility to see the affair through. Reluctantly, Timothy agreed to become chief executive of TV-am, with Jonathan replacing him at Aitken Hume. The IBA accepted the move.

Timothy knew precisely what he wanted to do almost as soon as he took over, and he did it. He summoned Angela and Anna to his office and fired them, giving as his reason their statements to the press on the day Peter resigned, technically in breach of their contracts. If the newspapers had been excited by Peter's departure, now they were ecstatic. Peter, after all, was only mildly famous, too aloof to be a popular hero, without an especially pretty face. Angela and Anna were two of the best-known television personalities in the land. The Famous Five, at a stroke, had been downgraded to the Trembling Three, wondering whether they would fall victim to Timothy's wrath.

Anna was supposed to be on holiday but early on Tuesday 19 April 1983, Timothy was driven by his chauffeur to her Brentford home, where he handed over a letter asking her to report to the office that day to be dismissed. Angela was due on duty anyway. Anna arrived first, at 9.30, to be told that she was being fired for having breached her contract. An hour later Angela was given the same message. By lunchtime both had left, looking shaken, after making their emotional but necessarily peremptory farewells. Neither would say much to reporters. Anna explained that her solicitor had advised her to keep quiet. Angela, asked for a quick word, replied: 'The quick word is no.'

But Michael Parkinson more than made up for their reticence. BLOODY SHABBY was the main front page headline on the next day's *Daily Express*. That derived from two colourful phrases he employed to characterise the sackings. They were 'a rotten, bloody shame', he raged. 'I think they have been shabbily treated. It's a disgrace. . . . I found it thoroughly nasty. I am very upset tonight about it all and I am considering what to do.'

Michael told the *Daily Mail* where *he* thought the axe should have fallen: 'I think some heads should roll at TV-am but they are not those of the presenters and the people who won the franchise. Anna and Angela, for too long, more than anyone else, have had to carry the butt of a badly-produced programme. It is convenient to blame them but only 10 to 15 per cent of the blame is theirs. There has been no leadership from the top and that's what makes me sick. . . . You can hardly blame the Aitkens. Those people I blame are the senior executives whose job it is to edit the programmes. The younger kids were given no lead at all and worked their socks off.'

When Parkinson flagged, his wife and co-presenter Mary took up the refrain: 'Michael is being an honest and honourable person in speaking out like this. He's doing just what I expected – to stand up and say what he feels. But I have never seen him more upset than he is today.'

It was all very well Michael considering what he should do next, but what the world waited breathlessly to know was what Timothy would do about him. The women, after all, had been dismissed for violating their contracts by talking to the press without permission. Now Michael had been even more outspoken. Logically, he should be the next to go. He was called in to see Timothy the next day. The press built up that meeting until it seemed as momentous as when Chamberlain went to see Hitler in Munich in 1938. Would he call out the Panzer divisions or would he sign a piece of paper promising no further excesses? The *Standard* was predicting the Panzers:

'Michael Parkinson was on his way to the TV-am studios today expecting to be sacked,' their main front-page story began. He said he had virtually made up his mind to leave and his weekend breakfast shows would be coming to an end.

'Tactically I am playing into their hands and I am just waiting for them to see what they are going to do. I am waiting for their reaction. ... I can't go this far down the line and let it run on.'

He said the Aitkens were shrewd businessmen who had successfully taken over the company. 'But if part of that takeover is picking off Anna and Angela in the way they have, do I have to keep quiet and collect the money, which I could do? That would involve a distasteful compromise. Or do I walk away from it and lose everything? I am prepared to do that.'

After criticising the 'Oxbridge school of journalism and philosophy' that had permeated the programme department ('They have firsts in English but can't actually read or write a running order'), Michael went on:

> If they [the Aitkens] had decided Angela and Anna did not figure in their plans, there was a proper and decent way of doing it. Give them the proper recompense for winning the franchise and for the two years they worked for nothing. They should be properly compensated and not made to look like naughty children and told not to talk to the press. It makes me cross. ... The game has fundamentally changed now. Someone has formed a new team, a new method of playing and I don't know if I'm interested at my time of life. So we will see.

And within hours we did. Primed with those fighting words, reporters waiting at Camden Lock, while Michael was talking to Timothy and Dick Marsh, expected to see him emerge with his head tucked under his arm. Instead, after a meeting lasting nearly four hours, he produced the equivalent of Chamberlain's piece of paper.

'I am staying,' he said. 'I am now much happier about the way the company is being run. ... There was a problem of communication and I have now communicated properly with Mr Aitken. ... The meeting started off aggressive and ended up much

189

more civilised. We both had one punch ready. He was going to sack me and I had my resignation prepared. It was all very frank and that is how I like it.'

Timothy had forgiven Michael his critical remarks of the last few days for a purely pragmatic reason: the figures showed him by far the most popular of the presenters and it was important to the company that he should stay. As a powerful lure, he offered Michael a seat on the board, saying that if he really wanted to criticise management he should be able to do something about it. After listening to Timothy's revelations on the way the company had hitherto been run, Michael accepted his offer. Timothy told him he agreed with his strictures about the programmes. He gave his assurance that Michael Deakin was being removed from any direct control over that area. Hilary Lawson, Parkinson's other *bête noire*, had already handed in his notice.

But if Timothy thought Parkinson's volte-face would reduce the emotional pressure, he had reckoned without the two women, who were attracting significant support from the press and public figures. The broadcaster Terry Wogan, attending a celebrity lunch with Angela, said: 'What a way to run a television station. ... The firing of Anna and Angela discredits the whole company.' The *Standard* ran a disapproving editorial: 'The real culprits are not the presenters or the company but the IBA. By handing the franchise to a consortium of superstars, rather than to one of its more experienced rivals, the IBA endorsed the idea that people who make news can also self-effacingly present it. They misjudged the morning mood of Britain. It seems unlikely that much can now be done to salvage that mistake.'

There was nothing self-effacing about the wounded victims, notably Angela. In an interview on BBC television she spoke of the 'dirty tricks department' at TV-am that had tried to sow discord among the presenters. She added that she thought she had been sacked not for speaking to the press but for 'reasons which are more unpleasant than that, but I can't go into them.'

Her agents, the Mark McCormack Organisation, were said to be soliciting offers of more than £50,000 for Angela's exclusive account of the unpleasantness behind the dismissals, but by Friday the offer had been withdrawn. She summoned a press conference – something that made Timothy so nervous he was considering taking out an injunction to stop her. But he need not have worried. Her burden was that although she had received 'eight substantial offers' for her tale, 'I believe it is morally and professionally wrong for me to abuse confidences of my employers.'

She did not let them get away with it quite that easily. 'There are people at TV-am who have acted in a shameful and disgraceful manner,' she said, 'and in the short term it might be very nice for me to, if you like, gain revenge. It would be easy to say a lot of things that are very damaging, but that would be destructive.'

Timothy was himself not doing much to lower the temperature. As part of his attempt to bring the gravity of the position home to the staff, hoping they would accept economies, he assaulted the record of the previous administration in yet more devastating terms than cousin Jonathan had employed. Having used that argument to persuade Michael Parkinson to stay, he wielded it again when he addressed the staff a few days after taking over, with Roger Frye in support to provide the financial details. An unsourced report in the *News of the World* the following Sunday read as though it derived from a transcript of that meeting.

'It was insane,' Timothy was quoted as saying. 'There were no contingency plans for failure. When I told the board that there should be some, they told me not to be silly.' He complained that he could not even sell the company cars because they were leased rather than owned outright.

'Even the pot plants are leased,' he asserted. 'We can't sell them. If I'd been given £15 million to start TV-am with I'd have run it from a tin shack. I'd have put £10 million in the bank and lived off the interest. And I'd have built a real studio on the back of the audience we would have been winning.'

He was even more outspoken when interviewed by Jean Rook in the *Daily Express* in May. 'We're finished with megastars and a mismanagement that solved every crisis by writing another cheque,' he declared.

My main job is to rebuild bridges and the relationships the Famous Five – fatal title, that, even without Enid Blyton – broke down. They were very arrogant. They were going to invent a new form of television, give a message to the nation and show everyone how to do it. What they couldn't understand was that they weren't a national mission. They were five people who ought to have been producing an interesting and entertaining programme.

He said they were unpopular with other stations in the ITV system because of their 'lordly insistence that they knew it all, when they knew so little that Angela Rippon and Anna Ford were shocked and shaken when I took a full hour to explain to them, when I sacked them, just where it had all gone wrong.

'And gone wrong so quickly for one simple reason. They believed their own publicity. They thought just being there would bring in the viewers.'

So incensed were they by these and other remarks by Timothy that Anna and Angela added libel actions to the claims they were making for compensation for their dismissal. Timothy maintained that they were technically entitled to nothing because they had, in his view, been dismissed for cause. He offered them £25,000 each. Anna wanted £137,500, most of it compensation for the 22 months of her contract left to run, and Angela's claim was broadly similar.

A corollary of the 'look-at-the-mess . . .' syndrome is that the people who served the old regime become irreversibly tainted from association with it, even if the failure was nothing to do with them. In the weeks following Timothy's takeover many people departed, less spectacularly than the two women but for the same essential reason – that Timothy wanted them out.

Tony Wakeling had his job pulled from under his feet when Timothy brought in Roger Frye as finance director. Derek Stevenson was replaced as sales director by Tony Vickers from Link Television (which Derek later joined). Derek had been faced with an impossible task, trying to sell advertising on a station whose viewers were dwindling to nothing. His differences with Timothy were as much personal and philosophical as to do with his competence. Timothy disagreed with Derek's stress on regional advertising, believing that the main sales appeal of TV-am was as a national network offering a national audience. By the autumn of 1983, Vickers was selling twice as much national as regional advertising – the reverse of the position in the spring and summer. As for the personal differences, Timothy found Derek petty. Soon after he arrived he reorganised the office arrangements, moving himself and Michael Deakin out of the programme area on the first floor to the business department on the ground floor. Derek had to move upstairs, and resented it bitterly, insisting that his furniture and pot plants should be moved with him. And he was angry when Timothy criticised him in front of his staff.

Geoff Smith, the director of operations, left in the summer. Jennie Bland hung on as a director until the autumn and then resigned, complaining that under Timothy she and other board members were being given inadequate information. Soon only Michael Deakin remained from the top stratum of the old regime. The IBA had let Timothy know that a change in the director of programmes would render the company so unrecognisable from the group that won the franchise that the Authority would be even more embarrassed by the turn of events than they already were. But Michael's authority was restricted. As Timothy cut his swathe through the old guard, the survivors bestowed a nickname on him: Pol Pot.

Timothy had tasks as urgent as clearing the stables, though less liberating. A formidable problem Jonathan had left for him to tackle was the renegotiation of union contracts. The simplified shift pattern would make sense only if the unions would agree to the lower take-home payments implicit in it. Timothy also wanted them to agree to forgo, or at least delay, a 9 per cent increase in pay due that year. His public pleas of poverty made an impression on the unions, as did the announcement that David Frost and Michael Parkinson had agreed to a pay cut of some 25 per cent, taking them back from six figures to five. (This was the cut they had discussed with Jonathan earlier.) Timothy's and Roger Frye's talks with the unions took place in the canal boat that Peter had bought for £18,000 and that the Aitkens produced as yet more evidence of waste. It was supposed to be a recreation centre for staff but was seldom used. It did, however, provide a venue for the union bargaining sessions that lasted eight days. At the end of it, the unions, by now convinced that the survival of the company really was at stake, had agreed to nearly everything Timothy asked.

On screen, Anne Diamond was recruited from the BBC to replace Lynda Berry as joint host with Nick Owen of 'Good

Morning Britain'. They settled into a likeable team,
strengthened later in the year by the arrival of the experienced
John Stapleton to handle serious interviews and introduce a
touch of gravity — carefully controlled — to a programme that
was beginning to get a little too lightweight in the attempt
to attract viewers. Together, the three were paid less than
the £112,500 David was getting before his voluntary salary
cut.

In June there occurred a widely reported incident that seemed
to symbolise better than anything the end of the Famous Five.
On Tuesday, 14 June, Lady Melchett, widow of the former
chairman of the British Steel Corporation and one of London's
leading society hostesses, threw a party at her home in Chelsea.
The guests included James Callaghan, the former Prime
Minister; Lord Hailsham, the Lord Chancellor; Lord
Weidenfeld, the publisher; and several members of Parliament,
among them Jonathan Aitken. Anna Ford was there with her
husband Mark. Seeing Jonathan chatting amiably in a group
on the far side of the room, she manoeuvred through the guests,
carrying a glass full of white wine which she threw at the side
of his face, soaking his jacket and shirt. Jonathan, trying to ignore
it, carried on talking. Her magnificent gesture would have been
meaningless had Anna not subsequently explained her motive
in moving detail:

'It was the only form of self-defence left to a woman when
she has been so monstrously treated,' she told the *Standard*.
'He had just ruined my life. He broke my contract and I lost
my job.' She had not known he was going to be at the party
but when she spotted him she could not resist the opportunity.

'It was a pretty good aim. I hit him absolutely full square.'

Jonathan's response was characteristically urbane.

As a father of three children under the age of three [two are twins],
I'm quite capable of remaining calm when dealing with
kindergarten theatricals. I was talking to somebody then, zoink —
some liquid hit me at the side of the face.

I turned around and saw her retreating. We did not speak to one
another. It wasn't a very mature way of dealing with her problems.
If at any stage she had wanted to talk about her dismissal, I have
been around.

The press adored the story. The cartoonists, who had not found much material at TV-am since Anna and Angela were fired two months earlier, suddenly found it a fertile field again. Articles about wine-throwing – how to do it, whom to do it to, what to wear – burst all over the features pages. In the *Daily Mail*, Auberon Waugh suggested wines with good throwing qualities, ones that travelled accurately.

The following October Anna settled her claim against TV-am for £43,000. Jonathan sought no deduction for his dry-cleaning bill. Angela held out longer and in January 1984, settled for £70,000.

Greg Dyke arrived at the beginning of May 1983, and at the end of that month introduced his new look. Daybreak was abolished entirely. Transmission now began at 6.25 and ended at 9.30, the IBA having allotted an extra quarter-hour of revenue-earning time. 'Good Morning Britain' ran for the whole three hours. Commercials came in larger batches with longer intervals between them, to make them less irritating. Each commercial break was preceded by an enticing foretaste of things to come in the programme, to dissuade people from switching off. There were competitions, pop videos and more show-business celebrities. Diana Dors, the buxom film star, did a series on dieting. The bingo numbers from the daily press were read out, so that people would not have to buy the papers.

In July, when the schools were on holiday, the new format began to get a response from viewers, and it came with extraordinary suddenness. With Frank Bough and Selina Scott taking holidays, the BBC were making do with substitute pairings, assuming that morning audiences would fall off in the summer at the same rate as those at other times of day. Dyke had other ideas. Research from Australia showed that summer breakfast audiences were boosted by children who did not have to get ready for school. So he kept his first team of presenters working through the school holidays and introduced children's programming for the last half hour of daily transmissions, featuring the cartoon characters, Roland Rat and Kevin the Gerbil. For adults, the holiday mood was sustained by taking an outside-broadcast unit (Timothy had bought a second-hand one cheap) to seaside resorts and devoting long segments to interviews from the beach, on pedal boats and the like. This made for easy viewing as well as being a good way of promoting the station.

The figures show what a remarkable effect these moves had. By mid-July Dyke had succeeded in boosting the peak quarter-hour viewing figure from 300,000 to half a million – still much too low. In the last week in July it doubled to a million and did not stop there. By mid-August it was up to nearly 1,750,000, with the BBC down to 1,200,000. When the schools reopened it dropped back briefly below the million but many of the friends won during the summer stayed loyal and by October the million mark was being exceeded regularly. This meant that Tony Vickers, the new sales director, could begin to charge economic rates for advertising time: and Greg Dyke could claim the extra £20,000 on his salary.

Inevitably, the summer miracle provoked jokes about the rat that saved the sinking ship, but Timothy could take those. He did not even mind too much when *The Times* wrote a sniffy editorial about it, headed A RAT TO THE RESCUE:

> When Mr Dyke rode to the rescue, there were fears that he would take an exclusively low road to recovery. This he has done. The IBA, no doubt, is mightily relieved that it is not faced with a bankrupt, no-hope television station. The story of Camden Lock has shown that the IBA has both failed in its statutory duty as the public's guardian of quality and that it has no influence, despite the power of the franchise, for shaping public taste.

Michael Deakin rushed to the station's defence, objection to the transmogrification of TV-am by the press into a rat. 'What rankles most about our new-found anthropomorphous identity is the growing realisation that both the means and the end of 'popular' and 'commercial success' are considered by our more erudite critics to be somehow intrinsically distasteful or discreditable. In our view they are neither. *Vivat* Rat!'

The weekend news was better, too. Michael Parkinson's Sunday show had never been as popular as the Saturday one. With Parkinson away guest presenters were engaged and in September David Frost took over on Sundays. Within a few weeks he had lifted the Sunday rating above a million for the first time, doing a lot for his self-esteem which, however well he had hidden the fact, had been badly bruised by his failure to pull the viewers in February.

The better ratings improved morale but were by no means the complete answer to TV-am's problem. Timothy and Roger Frye still needed more money to cover the prodigious cash outflow of the first half of the year. Income was improving now that advertising rates were up, but hand in hand with that they needed a severe cut in expenditure. The agreements with the unions contributed significantly and the IBA chipped in by agreeing to postpone levy payments due. The little house built next to the studio ('Peter's folly', as it was now being unkindly called) was let to a film producer for £6,500 a year unfurnished. There were plans to let the smaller of the two studios, seldom used, to other TV companies. By November, Frye had cut the first year's expenditure from a budgeted £21 million to £14.5 million.

That put them in better shape to seek more capital. The need was evident. Barclay's more than once threatened to cut their credit facility further. From time to time unsettling reports would appear in the press, suggesting doubts about the company's ability to pay wages and other bills. Jacob Rothschild, still smarting over the treatment meted out to Peter, had pulled out in the summer. Timothy calculated they needed another £4½ million to survive and began trawling for new investors. Several intriguing names were proposed. Ladbroke's, the bookmakers, were interested for a while. So, according to press speculation, was Robert Holmes a' Court, the Australian high-flyer who had taken over Lord Grade's Associated Communications Corporation. Another overseas name that arose was that of William Paley, the eighty-two-year-old American tycoon who helped found CBS, one of the three national TV and radio networks in the United States. Towards the end of October Timothy made a tentative approach to Robert Maxwell, head of the British Printing and Communications Corporation. Timothy asked him whether he would be interested in taking a 20 per cent share in the company for £2 million. Maxwell was hesitant but late on the afternoon of

Friday, 28 October – Timothy's thirty-ninth birthday – an alarming rumour began circulating in the City. It was said that TV-am had failed to raise the money it needed to survive and was likely to go under within days.

Soon the word reached newspaper financial desks who sought confirmation. 'You are the forty-third person to ask me about these rumours,' Michael Deakin told Philip Robinson of *The Times*. 'We are still in business and everyone is very calm.'

The following Monday Maxwell made a cash offer of £3 million for 30 per cent of the equity. This would have made him the largest single shareholder and given him an important measure of control. Timothy was able to turn him down, for his negotiations with an alternative source of finance were about to come to fruition. When discussing between themselves whom they might approach both he and Michael Deakin thought of Lord Matthews, chairman of Fleet Holdings, publishers of the *Express* newspapers and the *Daily Star*. They were delighted when Matthews agreed to take 20 per cent of TV-am for £2 million. Among the Aitkens' long-term ambitions has been to restore the family's newspaper interests and this link with the successor to the Beaverbrook empire might be a move in that direction. But it still left them with £2½ million to raise. Of this, £1½ million came from existing investors. In mid-November the final £1 million slotted into place when it was revealed that Kerry Packer, the chairman of Australian Consolidated Press, was buying 10 per cent of the equity and would have a seat on the board.

Packer, son of the legendary Sydney newspaper chief Sir Frank Packer and owner of a thriving Australian television network, was best known in Britain for his role in revolutionising cricket. By paying the world's best players large amounts of money to join his 'pirate' league he succeeded in altering the commercial structure of the game, to the dismay of the establishment diehards who preferred it as it was. An old friend of Rupert Murdoch, Packer shares Timothy's buccaneering approach to business – as does the other new major shareholder, Lord Matthews. Between them, these three now owned about half TV-am's equity.

By 1 February, 1984, the first anniversary of that heady

opening programme, just two of the original gang remained. David Frost seemed happy with his successful Sunday show, while Michael Deakin, clean-shaven now to give him less of a Mephistophelean aspect, sat warily two offices away from Timothy, still making jokes. Nowadays not many people came and sat companionably on his floor, swapping ideas. Michael Parkinson, reduced to doing only the Saturday show, now felt that he was not playing in the first division and was determined to try his luck in Australia on a more permanent basis, although he would remain a shareholder and director of TV-am. Robert Kee was still available for special assignments but no longer on the permanent strength. Anna Ford was doing interviews for the BBC's midday programme, 'Pebble Mill at One', while Angela Rippon had begun working in the United States as arts reporter for a Boston TV station. Peter Jay was hosting a weekly political programme for Channel 4 and had just endured another painful bout of front-page publicity over the allegation by his children's former nanny that he was the father of her four-year-old son.

Ratings were regularly level with the BBC's at nearly 1,300,000 but financially the company was still troubled, partly due to the continuing Equity/IPA dispute. Timothy warned the staff that more redundancies would be required – at least 40 and maybe as many as 90 if no further finance was forthcoming. The press was full of reports of crisis meetings and threats of imminent closure.

It had been a dreadful year for all of them. Perhaps they looked back at Peter's warnings during the run-up to the start of broadcasting: 'We'll be Murdoched,' he had cautioned. They had not precisely been Murdoched. But they had been Matthewsed and Packered and, above all, Aitkened, which they might feel amounted to much the same thing. Certainly they had succeeded in discrediting the United Artists concept of creative people coming together to run their own show – but then, as Michael Deakin would observe mordantly, the original United Artists had not worked out too well, either.

David Frost is not easily put down. He at least knew how to celebrate TV-am's first birthday. He presented each member of the staff with a quarter bottle of champagne. The toast: survival.

Dénouement

At the end of an Agatha Christie mystery, the detective often calls all the characters together in the drawing room, identifies the person responsible for the crime and explains the devilishly intricate reasoning that led the sleuth to that conclusion. In this case the question is: Who pushed Peter Jay? And was it truly, as Anna Ford maintained, a matter of treachery?

The first suspect we can eliminate is the mission to explain. It has the perfect alibi, having made no appearance at the scene of the crime. It cannot therefore be held responsible for TV-am's failure to attract any significant audience in the first month. The episode does not tell us whether the Jay-Birt thesis is viable in the morning because it was not tried. The programmes were inaccessible not because they contained too much boring explanation of the news but because the presenters and producers struck the wrong tone. They were slick, knowing and sometimes, sin of sins, patronising. That bit of Beethoven at the end of the Valentine's Day programme was a case in point, almost an apology for the shameful banter that had gone before, a bit of culture slipped in before anyone had time to switch off. It smacked of those crude attempts by worthy Victorians to improve the minds of the working man and woman.

Peter Jay believed it was the very absence of the mission to explain that caused the debacle. In a letter published in *The Times* on 4 October 1983 he wrote that he was haunted 'by our failure in the first few weeks even to try to make the kind of programmes we had talked about and the undeserved damage which this failure caused to the careers of many excellent and dedicated people – presenters, reporters, technicians, salesmen

200

and others – whose fault it absolutely was not.' His omission of programme-makers from the honour roll of the blameless indicates whose fault he thinks it was.

In his letter Peter reaffirmed his belief in the Jay-Birt thesis as a reaction against the 'green eye-shade and suede jacket establishment of television news and current affairs, with its twin inheritance of reflexes from the Gateshead news desk and from Hollywood'. This provoked a stinging reply from Dennis Forman, the chairman of Granada Television. It showed the resentment Peter's theories had engendered among TV professionals and why many were not sorry to see him fall on his face.

> It is interesting to learn that so many of my old broadcasting colleagues were secretly addicted to green eye-shades and suede jackets, for one never saw them wearing these articles in public. Similarly it is surprising to learn of their clandestine visits to Gateshead, for these were never mentioned in open conversation. . . . The production of programmes was their first priority and, since it is a very demanding occupation, they had time to do little else, even to write one single magniloquent thesis about the profession in which they were, and in all cases still are, successful practitioners.

Another response to Peter's letter bore more relation to the mystery we are gathered here to solve. Hilary Lawson agreed with him that the reason for the failure was not the programme philosophy but the lack of any attempt to put it into effect.

> It would be sad and damaging if programme-makers drew the lesson from TV-am that populism pays. The task in broadcasting is to produce popular programmes that are more than pap. The mark one version may have failed to be popular enough but with more time (Roland Rat was, ironically, one of its creations) and with the right organisation it might have succeeded. Just as it is a mistake to write off the mission to explain on account of those early weeks, it is also a mistake to assume that the race for ratings need force one to programme bingo in the place of news.

That takes us into the unending argument about lowest-common-denominator journalism. It will not be resolved here.

Many would maintain that we are in any case chasing a classic Agatha Christie red herring by examining the reasons for the failure of the programmes. A more conspiratorial theory has it that the Aitkens were always determined to seize control of the company, however successful it proved. The programme failure made it easy for the Aitkens to do the deed quickly, this theory goes – and their motive for such haste was that they feared the ratings would soon improve and their chance be lost.

Several factors fit that interpretation, but since they also lend themselves to alternative explanations they are far from conclusive. Timothy Aitken's ferocious criticism of the management of the company in 1982 could have been aimed at undermining Peter. Equally it could have been the response of an alert and responsible director to the failures Peter confessed to in his 'mistakes of May' document.

Jonathan's motives are difficult to discern, concealed beneath a politician's veneer of politeness and *bonhomie*. Those who single him out as the chief conspirator point specifically to his action in January 1983, when he took the lead in abandoning the search for a managing director. Only a few months earlier he and Timothy had been the most vociferous advocates of such an appointment. Did he foresee that things were going to go dreadfully wrong? Was he merely allowing Peter enough rope to hang himself, leaving the Aitkens free to move in? Perhaps, yet the arguments Jonathan deployed at the time remain persuasive. It was a crucial appointment, not one to be rushed. No suitable candidate had materialised during the three-month search. The wrong appointment would be worse than no appointment. And it would have been a dreadful extra burden on the appointee to plunge him or her into the chaos surrounding the start of broadcasting.

The 'Aitken plot' scenario implicitly involves Michael Deakin, whose letter to his old friend Jonathan set in motion the final stage of the power struggle and was what Anna Ford had in mind when she alleged treachery. The letter was provoked by Michael's frustration at the ratings failure and his personal clashes with the presenters but it is hard to see it as a consciously treacherous act. Treachery requires the calculation of some personal gain and Michael must have recognised that Peter's departure would be against his own best interest. A new chief executive, less bound than Peter by friendship with Michael's chief patron David Frost, would be likely to detect the need for changes at the top of the programme department. Michael did not, when he wrote the letter, know of Peter's renewed approach to Mike Townson – indeed, he could not have known, for the approach had not then been made.

As for the suggestion of a scheme worked out with the Aitkens in advance, the letter is evidence against rather than for it. Had there been such a plot, Michael would not have had to write to Jonathan to provoke him to act, because the need for action would already have been understood. And any 'treachery' was poorly rewarded, for although Michael has kept his title as director of programmes, his power has been greatly reduced. If the IBA had not insisted that he stay in place, it is conceivable that he would have gone the way of Anna Ford, Angela Rippon, Geoff Smith, Derek Stevenson and the others.

So what about the presenters? Were they to blame? Here we are not looking for a conscious plot against Peter, their greatest supporter, but for actions by them that inadvertently brought about his fall. It was Anna Ford and her 'flu that gave rise to the Deakin letter. Therefore she bears some responsibility for the timing, at least, of the offence we are investigating. And to the extent that the ambiguity of their position vis-à-vis the programmes department was never properly resolved, the presenters contributed to the confusion, leading to the poor programmes that in turn provided the pretext for the Aitken *putsch*. They were supposed to have 'input' but no decision-making powers. In practice nobody could work out what that meant. Television, like the press, is an industry with few objective standards of quality and scope for endless disagreement about the right way to do it. Much of it comes from the fingertips and different sets of fingertips give off contrasting vibrations. With no strong leader at the helm, the Famous Five formula was a recipe for chaos. Yet it is unfair to blame the presenters, except in so far as they were taken in by the unrealistic bill of goods David Frost and Peter Jay sold them in 1980.

David Frost has been high on many people's lists of suspects because he has a record. He was involved in the great London

Weekend job in the late 1960s, bearing striking similarities to this débacle. Glittering names were brought together for the nefarious purpose of making a new and exciting form of television, which in the end few people wanted to watch. Many reputations were damaged. Yet in the intervening years David had largely purged himself of any guilt by performing good, popular and profitable works across the globe. It is a firm principle of British justice that a chap's misguided past should not be taken into account when deciding the verdict.

There remains a separate charge against David, that his early morning on-screen deportment deterred viewers from watching the first month's programmes. He has secured a convincing alibi for this by proving successful in his more recent Sunday morning appearances.

The IBA have been accused of complicity by having awarded the franchise to the most star-studded consortium, giving too little thought to the quality of the people behind the cameras. Public bodies composed of random representatives of the Great and the Good are easy targets for criticism and mockery. It is not hard to raise a smile at the thought of an unworldly Scots cleric grappling with complex technical and artistic matters outside his experience. The system for allocating commercial television and radio franchises is peculiar and irrational, yet it usually works. Would any of the other contenders have done better? The runners-up were the group headed by Christopher Chataway and Harold Evans. Evans's subsequent experience as editor of *The Times* proved that not everything he touches turns to gold. As for ITN, their difficult first year with Channel 4 news does not suggest they would have done well with breakfast television. There is insufficient evidence to pin all the blame on the IBA.

That leaves us with one suspect, Peter Jay: and here we have the advantage of a confession. In the letter to *The Times* from which I quoted earlier he wrote:

> The reasons why at TV-am, instead of Jacobson's vigorous and intelligent *Mirror* we got the *Guardian* without the flair, are too tedious – and too painful – to explain here. Suffice it to say that the captain of the ship should, did accept and does accept the blame – for being so preoccupied with the business, sales and

operations part of his ship that he failed till battle was joined to realize sufficiently what was happening (and not happening) in programmes.

Seasoned readers of detective fiction will know that a confession is not conclusive proof of guilt. False confessions may be made for a number of reasons – under coercion by interrogators, to protect the truly guilty or simply because it seems the upright thing to do. Although the third of these motives could apply here my inclination would be to accept the confession – qualified though it is – at face value, and to look no further for the culprit.

Peter's mistake was that he believed his evident talent as a clear-headed and original writer about politics and economics, his confident grasp of difficult issues, qualified him as a man of affairs, able to translate his ideas into action. His gallant and sincere attempt to turn TV-am into a model company run by and for its workers, where reason and harmony prevailed – a kind of commercial equivalent of Welwyn Garden City – was doomed to failure. You cannot introduce new rules if everyone else is playing by the old ones. If you try you will sooner or later be sent off the field.

Ladies and gentlemen, thank you for your attention and patience. The butler has some cold cuts ready in the dining room and the chauffeur will ferry you to the station for the London train. I hope you will ponder the events I have described and profit from the lessons they contain, to avoid making the same mistakes in future; but I doubt it.

Index

Index

Index

Stevenson, Derek 30, 35, 96, 98, 99, 102, 121–2, 124, 137–8, 143–4, 152–3, 173, 180–1, 192–3, 203
Stigwood, Robert (and org.) 10, 141, 150, 160, 161 165
Stocken, Oliver 54
Stoppard, Dr. Miriam 59
Strauss, Norman 42, 43, 44, 50
Summers, Sue 76
Sun 15, 186
Sunday Express 129
Sunday Telegraph 10
Sunday Times 2, 57, 63, 140, 141, 177

Tebbit, Norman 27, 131, 133
Thames Television 32, 37, 38, 57, 58, 75, 79, 84, 96
Thatcher, Carol 186
"That's Life" 8, 19, 52, 65, 84
"This Week" 37, 38, 39
Thomas, George 88
Thomson, Lord 15–16, 17, 26, 63, 64, 65, 66, 67, 78, 105, 133, 151, 155, 162, 179–80
Time (magazine) 2, 171
Times 42–3, 46–7, 75, 91–2, 123, 128, 131, 164, 196, 198, 200, 204; Jay on staff 3–7, 14, 67
Timpson, John 128
"Today" (BBC) 46, 77, 128
"Today" (NBC) 92
Tonypandy, Viscount, *see* Thomas
Townson, Mike 32–4, 37, 58, 84, 152, 155–9, 163, 183, 203
Trethowan, Sir Ian 93
Trilion Video 58
Trumpington, Baroness 59
TV-am 51: franchise bid 12–14, 51–5, 58–9, 64–9, 73–5; early days 84–8, 93, 96, 98–100; staff 21, 24, 109–13, 120; and ITN 20, 55, 89–91; bldg. 19, 23–4,

80–4, 96, 104, 107–8, 113, 167–8, 181; problems 94, 114–5, 118, 120–8; on air 27–8, 129–40, 144, 148–50; Aitken coup 160–92, 195; TV-am News 163
"TV Eye" 32, 37, 58

United Artists 8, 13, 30, 50, 97, 199
United Medical Ent. 47, 49
UPITN 91

Vickers, Tony 192, 196

Wakeling, Tony 94, 97, 98, 99, 102, 103, 118, 122 192
Wallace, Marjorie 177
Warnock, Mary 64, 75
Watt, David 42
Waugh, Auberon 195
"Weekend World" 4, 5, 53
Weidenfeld, Lord 194
Weinstock, Lord 16–17, 25
Whicker, Alan 31, 59
Whitelaw, Viscount 179
Wilcox, Desmond 51
Williams, Shirley 2
Wilson, Charles 32, 150, 154
Wilson, Sir Harold 12, 41, 48
Wiltshiers (builders) 84
Wintour, Ch. 50, 59, 61, 164, 186
Wogan, Terry 190
Wooler, Michael 57, 75
Worcester, Robert 60
"World in Action" 38
Wyatt, Woodrow 16

"Yorkshire Mafia" 21, 115, 116
Yorkshire TV 4, 10, 12, 21, 72, 75, 97, 115; Deakin and Frost at 8, 31, 50, 87, 110; breakfast tv experiment 40–1, 44
Young, Sir Brian 44–5, 71–3, 75

211

16